T0275722

Solar Photovoltaic Technology Production

Solar Photovoltaic Technology Production

Potential Environmental Impacts and Implications for Governance

Senthilarasu Sundaram, David Benson, and
Tapas Kumar Mallick
Environment and Sustainability Institute
University of Exeter, Penryn
United Kingdom

AMSTERDAM • BOSTON • HEIDELBERG • LONDON
NEW YORK • OXFORD • PARIS • SAN DIEGO
SAN FRANCISCO • SINGAPORE • SYDNEY • TOKYO
Academic Press is an imprint of Elsevier

Academic Press is an imprint of Elsevier
125 London Wall, London EC2Y 5AS, UK
525 B Street, Suite 1800, San Diego, CA 92101-4495, USA
50 Hampshire Street, 5th Floor, Cambridge, MA 02139, USA
The Boulevard, Langford Lane, Kidlington, Oxford OX5 1GB, UK

Library of Congress Cataloging-in-Publication Data
A catalog record for this book is available from the Library of Congress

British Library Cataloguing-in-Publication Data
A catalogue record for this book is available from the British Library

ISBN: 978-0-12-802953-4

For information on all Academic Press publications
visit our website at https://www.elsevier.com/

Working together
to grow libraries in
developing countries

www.elsevier.com • www.bookaid.org

Publisher: Joe Hayton
Acquisition Editor: Lisa Reading
Editorial Project Manager: Natasha Welford
Production Project Manager: Anusha Sambamoorthy
Cover Designer: Greg Harris

Typeset by SPi Global, India

Contents

Preface

Population, poverty, and climate change are three of the most serious issues faced by our society in this century. The global increase in population has created an unequal distribution of natural resources. Industrialization and modernization have brought many changes to human lives. Energy is playing a pivotal role in urbanization and modernization. Developing countries are heavily urbanizing which leads to severe problems in power distribution and environmental challenges. The energy challenge has to be sustainable, and a greener approach to reducing carbon emissions and greenhouse gas (GHG) emissions is required. Using renewable resources in a wise manner can bring a massive change in energy production, which can contribute to improving environmental resilience and energy security. The green approach in the industrial process of renewable resources-based energy production has to be implemented at all levels of the production chain. Therefore, in this book we analyze materials used in generating solar electricity via photovoltaic (PV) cells, modules, and their industrial production. We also discuss the materials used in PV industries, discuss their lifecycle energy analysis, and suggest associated policies for governing them. This book gives a comprehensive introduction to materials mining, processing, and governmental policies of the PV process for the researchers, manufacturers, and the regulatory bodies.

We would like to express our appreciation and acknowledge a wide range of people, including our solar-group members, especially Dr. Sourav Khanna and Mr. Prabhakaran Selvaraj for their involvement in collecting data and sketching the necessary figures. We would also like to thank Linda Langstaff (Environment and Sustainability Institute, University of Exeter) for assistance in producing Chapters 4 and 5.

Senthilarasu Sundaram
David Benson
Tapas Kumar Mallick
Penryn, 2016

Chapter 1

Introduction

1.1 INTRODUCTION

The Industrial Revolution has brought many changes to our lives. Urbanization is one such change feeding the modern economy and social growth. Modernization and urbanization has its worst effects on developing countries, especially in growing economies like India and China who are starting to taste the bitter sweetness of modernization in terms of natural resource loss, increasing e-waste, poor air quality in many places, water scarcity and so on. Urban and modern world needs revolve around comforting surroundings, and controlling the environment through energy utilization. Our energy production majorly depends on fossil fuel, which produces significant amounts of carbon dioxide (CO_2) and other greenhouse gas (GHG) emissions. Fast-paced urbanization and industrialization of developing countries have triggered the fossil fuel demand since the 1970s. The increasing volume of carbon emissions and GHG emissions has become a catalyst for climate changes, global warming, and social welfare related issues. Developed countries and developing countries are very particular about controlling their emission levels and taking control measures. Environment and health issues related to carbon emissions and GHGs has impacted the way we look at energy generation.

Apart from fossil fuel based energy, the last two decades have identified and explored lots of alternate energy resources. Renewable energy resources are one of the best ways to deal with the issue. Renewable energy installation and usage trends tend to grow rapidly with a strong awareness of energy consumption and utilization. These increasing trends are not only associated with carbon emission awareness, but are also associated with the drop in oil prices, sustainable energy generation, and improvements in energy efficiency. The deployment of renewable energy is playing a key role in addressing the most important issues, such as climate change, creating new economic horizons, social welfare, and providing energy access to larger populations. In 2013, it was estimated that 19.1% of global energy consumption came from renewable energy resources. In terms of sector separation, the heating sector grew at a steady pace, there was an increase in biofuels, and there was a larger increase in the power sector, which is governed by solar photovoltaic (PV), wind, and hydropower.

Total annual energy consumption has been increased tremendously; as per predictions, energy production and consumption were pitched at 518.546 and

Solar Photovoltaic Technology Production. http://dx.doi.org/10.1016/B978-0-12-802953-4.00001-9

520.272 quadrillion Btu, respectively in 2011. This has been increased to 537.266 and 524.076 quadrillion Btu respectively, in 2012. In order to meet the energy demand, renewable energy sources, such as solar power and wind, water, tidal, and geothermal energy, have been used in conjunction with traditional energy resources. Solar energy is the most abundant and everlasting energy resource available on earth, with 1.73×10^{14} kW of solar energy continuously striking the Earth. It can be tapped through several ways. PV technology is one way to convert solar energy into electricity. It has been viewed as an alternative technology for the past couple of decades. At present, PV is contributing up to 1% of the total global energy consumption, and is expected to increase to ~16% by 2050.

PV/renewable energy firm sector growth has been driven by different factors, including sustainable energy generation, low cost, low carbon emissions, and long-term stability, and is supported by government policies. In many countries, renewables are broadly competitive with conventional energy sources. At the same time, growth continues to be tempered by subsidies to fossil fuels and nuclear power, particularly in developing countries (Renewables 2015 Global Status Report, 2015). This book aims to discuss the environmental impact parameters of PV industries, and its governance regulations. It will discuss, in detail, the PV industrial technological process, toxic material handling, and impacts on the environment and health. The associated policies and recommendations for material handling, processing, and generation will also be discussed in detail.

1.2 BRIEF HISTORICAL OVERVIEW AND CURRENT STATE OF THE PV SECTOR

Solar PV is considered to be a low risk investment due to its predictable output, and its technical maturity over the years. Global PV installations are expected to reach a cumulative capacity of 200 GW; it is comparable to 30 combined coal or nuclear power plants. However, two decades ago, the electricity generation of PV cells was only for small-scale applications. Until 2000, the cumulative PV deployment was just 0.26 GW (Branker et al., 2011; http://www.iea.org/, 2014). However, at the end of 2014, it reached 177 GW (http://www.iea.org/, 2014). The higher appetite for PV related electricity generation has resulted in the low cost production of electricity from PV. It is expected that it will become a mainstream source of electricity by 2050 (http://www.iea.org/, 2014). China is a fast growing PV industry; it stands second to the United States in PV installation (http://www.iea.org/, 2014). Since the last decade, Germany has remained the leading country in terms of PV installation (http://www.iea.org/, 2014). Currently, China, with its rapid growth, has become the world's largest producer of PV power. Fig. 1.1 shows the solar PV global capacity from 2004 to 2014.

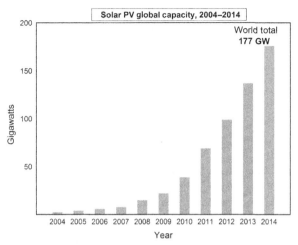

FIG. 1.1   Solar PV global installation capacity from 2004 to 2014 (Renewables 2015 Global Status Report, 2015).

Global solar PV installation has come from different technologies, such as silicon technology, and thin film and concentrated solar PV. However, crystalline silicon (c-Si) technology has more than 85% of the share in the present PV industry, through cells and modules based on poly, mono, and multicrystalline wafer technology. The sustainable growth of the PV industry and market is phenomenal, with a substantial surge, on average over 40%, recorded globally during the last decade. The silicon solar cell industry mainly depends on the semiconducting electronic industry for its feedstock. Thin film solar cells have created a niche in building integrated markets and window based applications. Third generation solar cells are aiming to establish a market within indoor applications. High/reasonable conversion efficiency of solar cells and good performance stability (25 years) are two measures for any solar technology coming into the market. Silicon and thin film solar PV technologies have proved their potential, and have created a substantial market size. In terms of technology development, the PV cells are generally divided into three generations. First generation solar cells are c-Si solar cells. The monocrystalline silicon cell is the most commonly used cell. The record efficiency of 27.6% is achieved by Amonix under 92 suns, whereas the highest efficiency, under 1 sun, is achieved by Sun Power which is 25.0%. Another c-Si solar cell is the multicrystalline cell, which can be manufactured with much simpler and cheaper techniques than those of the monocrystalline cell, but it has less efficiency. The highest achieved efficiency is 21.3% by Trina Solar. Second generation solar cells are thin film based cells, which are cheaper than conventional silicon cells. Their highest achieved efficiencies are 23.3% under 14.7 suns by NREL, and 22.3% under 1 sun by Solar Frontier. Third generation solar cells are an emerging technology, which is based on organic-inorganic hybrid solar cells. In the past few years, research on third generation cells has progressed rapidly. Other emerging solar cells are multijunction cells, which also come under third generation cells. These are the most efficient solar cells to date, but they are expensive. Their highest achieved efficiencies are 46% under 508 suns by Fraunhofer ISE using 4 junctions, and 38.8% under 1 sun by Boeing Spectrolab using 5 junctions. Thus, solar PV technology has progressed a lot. In 1977, the price for a c-Si cell was around $77 per watt; currently, the average prices are $0.335 and $0.536, respectively, for a cell and module of monocrystalline silicon. In spite of the progress in this technology, the key hurdles in PV installation remain, such as the uncertain availability of solar radiation, the storage of electricity for night hours, dirt accumulation on the PV panels, and so on. Apart from the technology hurdles, implementation, execution, and governmental support are the key factors that drive the market.

1.3   AIMS AND OBJECTIVES OF THE BOOK

This book aims to address a significant gap in the current literature on the solar PV sector. While, as discussed previously, multiple studies have examined specific technologies, and the impacts and challenges associated with this industry, there has been very limited discussion of the implications for their governance. In this respect, the book is truly interdisciplinary in that it combines a natural science analysis of the positive and negative impacts associated with different PV technologies with a social science perspective on developing potential governance options.

Meeting this aim has entailed the pursuance of four main objectives. Firstly, the book seeks to outline the major emerging solar PV technologies in order to provide context to the study. Secondly, it seeks to identify some of the main environmental impacts associated with these technologies, with an emphasis on threats to human health and safety. Thirdly, it seeks to globally map governance responses to the development of the PV sector, and its resultant environmental effects by comparing policy approaches in several leading states. Finally, the book seeks to explore the capacity of global governance to support sustainable development of the industry by expanding solar PV manufacturing and installation, while reducing impacts.

1.4   STRUCTURE OF THE BOOK

This book is structured around five chapters. Chapter 2 is about the different PV technologies, their industry status, and production details. Chapter 3 examines the potential impact of solar energy technologies. While solar PV panels are more environmentally clean than fossil fuel energy through their lower GHG emissions, materials employed in their production involve a number of environmental, health, and safety impacts, particularly from the hazardous substances used. This chapter consequently discusses different solar PV technologies, and how they are manufactured. It then sums up the potential environmental, health, and safety impacts of these various technologies, before examining what constitutes an acceptable limit for exposure to these chemicals and hazardous air emissions. The chapter also identifies other potential impacts, including the effects on the environment from solar power installation. Chapter 4 examines how these impacts are being governed globally. One key point developed in this chapter is that, while the solar PV industry has evolved rapidly in the last decade, approaches to governing its development and potential environmental impacts differ between national jurisdictions. Governance, which denotes the process of steering the actions of different societal actors (Pierre and Peters, 2000), can be measured in many different ways, but here we focus on a policy instruments approach (Jordan et al., 2012) in order to compare between five leading national contexts, namely: the United States, the United Kingdom, Germany, India, and China. These countries were chosen because they are significant manufacturers of solar PV technologies and/or installers of solar power. They also collectively dominate economic activity globally. National policy responses to promoting the development of the solar PV industry and controlling impacts are examined via a focus on the adoption of "command and control" regulations; market-based instruments, such as taxes and subsidies; voluntary agreements with the industry; and the provision of information, or informational instruments, by governments. Our comparative analysis shows that, while these states have done much to encourage expansion of the solar sector through a variety of policy mechanisms, the governance of associated impacts varies; thereby, suggesting the need for more harmonized responses globally.

Chapter 5 develops this point by exploring the potential for more effective global governance in the solar PV sector. The central theme developed by this chapter is that, due to the combination of the rapid globalization of the industry and the limited global institutional framework for governing its impacts, alternative governance solutions should be discussed. To initiate such a debate, the chapter argues that, rather than introducing prescriptive, top-down global institutions, industry-led voluntary approaches may be more appropriate in relation to the transnational nature of this sector, where the production of these technologies, and their utilization, is increasingly occurring across a globalized space. Here, we examine existing voluntary industry approaches to determine best practices, with a focus on corporate social responsibility, environmental management systems, lifecycle assessment of products, and extended producer responsibility. This chapter argues that these mechanisms offer a significant scope for integrating sustainability across the global solar PV sector, while additionally supporting the United Nation's sustainable development goals (UN, 2015). In this respect, we forward a set of normative sustainability principles, derived from preexisting industry guidelines and our analysis, to provide a linkage to these goals. It is hoped that they can then inform future governance on a global scale.

Chapter 6 then returns to our original aims and objectives to sum up our analysis. This book also places its findings within the context of available knowledge to show how it adds to the academic literature. Finally, it discusses areas of future research. One area identified for further in-depth investigation is how potential toxicological effects can be managed through more efficient waste management. Another research priority discussed in this chapter is the need for academics to work more closely with industry and policy actors in shaping the future governance of solar PV technologies in order to ensure their sustainability.

REFERENCES

Branker, K., Pathak, M.J.M., Pearce, J.M., 2011. Renew. Sust. Energ. Rev. 15, 4470–4482.
http://www.iea.org/publications/freepublications/publication/
TechnologyRoadmapSolarPhotovoltaicEnergy_2014edition.pdf, 2014.
Jordan, A., Benson, D., Wurzel, R.K.W., Zito, A.R., 2012. Environmental policy: governing by multiple policy instruments? In: Richardson, J. (Ed.), Constructing a Policy-Making State? Policy Dynamics in the European Union. Oxford University Press, Oxford.
Pierre, J., Peters, B.G., 2000. Governance, Politics and the State. Palgrave Macmillan, Basingstoke.
Renewables 2015 Global Status Report, 2015. Renewable Energy Policy Network for the 21st Century.
United Nations (UN), 2015. Transforming Our World: The 2030 Agenda for Sustainable Development. UN, New York. http://sustainabledevelopment.un.org/post2015/transformingourworld.

Chapter 2

Overview of the PV Industry and Different Technologies

2.1 INTRODUCTION

The majority of the (about 80%) world's energy needs are fulfilled by fossil fuels (Müller-Fürstenberger and Wagner, 2007; Luna-Rubio et al., 2012). Coal is sharing a major portion of the electricity production, however, it is projected that 40% of electricity generation from coal in 2008 will decline to 37% of electricity generation by 2035 (Sahu, 2015). However, the world energy consumption will be 50% more by 2030 (Sahu, 2015). The 45–50% of world energy consumption is by large-scale industries in the developing countries (Sahu, 2015). The different organizations such as the US Energy Information Administration (EIA), the International Atomic Energy Agency (IAEA), the International Energy Agency (IEA), and the World Energy Council (WEC) have given their future energy demands by different projections for 2020, 2030, and 2050 (Suganthi and Samuel, 2012). Replacing the fossil fuels with clean energy has paved the way forward to address the environmental challenges. The clean energy technologies are capable of meeting world energy demands as well as reduce the environmental challenges, especially global warming. Photovoltaic (PV) technology has been one of the fastest growing clean energy industries in the last decade.

PV industry is a fast growing industry with annual growth rate of 44%. The PV module production has also been increased to meet the current market. China and Taiwan have increased their PV installation compared to European countries. Si-wafer based PV technology accounted for about 92% of the total production in 2014. The share of multicrystalline technology is now about 56% of total production, whereas thin film (TF) technologies have a market share of 9% of the total annual production (Masson et al., 2012, 2013). The third generation solar cells are not making it into the market yet. There has been an increasing trend in achieving low carbon emission technology to produce electricity throughout the world. This is due to the concerns about greenhouse gas (GHG) emission from various industries, carbon footprint contribution through different human activities, and climatic changes.

The solar power generation industry is associated with health and environmental issues like other industries. However, it can be assured of low carbon emission from its life cycle (LC) perspective. The LC emissions per

Solar Photovoltaic Technology Production. http://dx.doi.org/10.1016/B978-0-12-802953-4.00002-0

kilowatt-hour are rather low, that is, 29.4–101.8 gCO_{2eq} (Barthelmie and Pryor, 2014) which is just 8.7% of coal-fired power generation (Weisser, 2007). The LC emission from the PV is mainly in its material, manufacturing, and end life stages. The PV panels are made up of solar cells as their main component and associated components. The manufacturing of solar cells is a more energy-consuming process in the whole process. The upstream process would take about 60–70% energy and possible GHG emission (NREL, 2015a).

The solar PV technologies have been classically divided into three broader categories, such as (i) first generation solar cells, (ii) second generation solar cells, and (iii) third generation solar cells. First generation solar cells are mainly focusing on crystalline silicon (c-Si) technology which has a more than 85% share in the present PV industry through the cells and modules based on poly, mono, and multicrystalline wafer technology. The second generation solar cells are mainly to reduce the materials consumption and explore new materials. This generation is based on TF materials such as Cu (In, Ga) Se_2, CdTe, CIS, and amorphous silicon (a-Si). The third generation solar cells are aimed at reducing the manufacturing cost and using environmental friendly materials for solar energy harnessing. The third generation is based on dye sensitized solar cell (DSSC), organic solar cells, and nanostructured solar cells (only applied to low-cost manufacturing involving in the solar cell making process).

2.2   FIRST GENERATION PV TECHNOLOGIES

The first generation PV technologies are based on the silicon material. Silicon is one of the earth's abundant materials with suitable band gap of 1.1 eV to harness solar energy. The silicon PV industry has a feedstock support from the Si-based electronics industry in its initial stage of development. The purity of the silicon decides the efficiency and performance of the solar technology. Eventually, higher conversion efficiency for standard size modules (average $\eta > 16$–18%) and extremely good performance stability (more than 25 years) are two essential requirements for any technology to successfully demonstrate its potential for market (Chopra et al., 2004; Jäger-Waldau, 2004). This is a mature technology with well-developed mass production in GW-scale. There are different qualities of silicon used in solar cell manufacturing. These are mainly classified into three different types of silicon PV modules used, specifically single-crystal, polycrystalline, and silicon film deposited on low-cost substrates.

2.2.1   Single-Crystal Silicon Solar Cells

Monocrystalline silicon solar cells are the most popular and oldest technology made from pure silicon on thin wafers of silicon. Monocrystalline silicon is made up of ordered crystal structures, with each atom ideally in its predetermined position. Single-crystalline silicon wafers are manufactured through slow and carefully controlled conditions. This makes them one of the most expensive types of solar cells with higher conversion efficiency of 25% (NREL, 2015b). Apart from the efficiency and stability of modules, c-Si offers a low module manufacturing cost at around $0.38/W. Monocrystalline wafer capacity was around 16.6 GW of 24.4% market capacity in 2015 (Power and Energy Industry of IHS Technology, 2015).

When compared to other silicon solar cells, monocrystalline solar cell modules have 4–8% higher power output of the same size module and lower balance of system cost (Power and Energy Industry of IHS Technology, 2015). The demands for the monocrystalline silicon solar modules are for the rooftop applications in Japan and United States due to their higher energy output per construction area. In these countries, total installation was about 10 GW in 2014 and expected to rise in the coming years. The manufacturing of monocrystalline systems is mainly from Taiwan PV companies. But the real obstacles are coming through the multicrystalline solar cells marketing sector which has a 75% of market share.

The monocrystalline Si solar cells are manufactured in different architectures to improve efficiency and stability. The common methods to make silicon solar cells are more or less the same, apart from the end purity of the silicon. The sequence of c-Si solar cell production from silicon wafer is shown in Fig. 2.1.

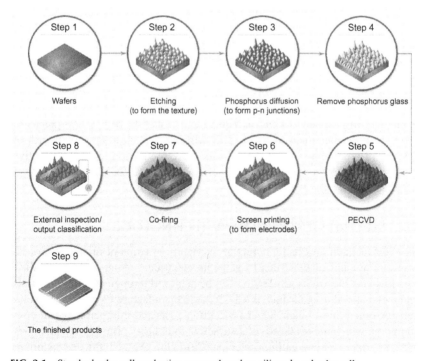

FIG. 2.1 Standard solar cell production process based on silicon based solar cells.

This is the most common method of manufacturing silicon solar cells. The fabrication of c-Si solar cells is starting with a clean silicon wafer, usually 300 μm thick. The cleaning or damage removal is vital in silicon solar cell growth to avoid recombination issues. The wafers will be coated with an antireflection coating (ARC). For a p-type c-Si substrate, an n-type top layer, while for an n-type c-Si substrate, a p-type top layer acts as emitter through a thermal diffusion. This will be followed by the edge isolation to create an electrical pathway and thin dielectric coating at the front and back of the wafers to passivate surface defects. The electrical connections will be made through suitable metal electrodes by different methods.

2.2.2 Polycrystalline Silicon Solar Cells

The cost effective methods to produce silicon solar cells were leading to polycrystalline silicon solar cells in the 1980s to use the silicon waste from the electronics industries. However, the conversion efficiency of the polysilicon solar cells was quite low, around 13% in laboratory cells (2 cm^2) which was not attracting many investors for this technology (Fally and Chabot, 1986). In fact, the demand for PV modules and the need for low-cost PV options in the last decade had stretched these advantages to the limit and had exposed some inherent disadvantages of c-Si technology, such as the scarcity of feedstock material, costly processing of materials and device fabrication steps, as well as the inability for monolithic interconnections (Upadhyaya et al., 2015). It has created huge interest in finding alternative materials and processes to replace c-Si PV. In 1990, with the announcement of the laboratory scale efficiency of 35% (in areas of 5 mm^2), the polycrystalline silicon solar cells manufacturing technology became interesting to investors (Braga et al., 2008). There is an estimation of 130 thousand metric tons (kilometric tons or KMT) of polysilicon manufacturing capacity, which is equivalent to 25 GW of c-Si PV panels in 2015–16 (Greentechmedia, 2014). This will increase up to 85 GWs worth of c-Si panel production. This is due to the easy production and cost effective methods involved to produce highly efficient (21.5% for 242.74 cm^2) solar panels (Green et al., 2015).

2.3 SECOND GENERATION PV TECHNOLOGIES

Si-wafer technologies had a potential restriction in terms of materials processing and amount of materials to generate electricity, along with its rigidity. The low-cost electricity generation can be achieved through cheaper materials and less material usage, which is a TF route to manufacture PV panels. TF solar cells emerged due to their lower production costs and minimal material consumption, which makes these cells attractive to industry. The key technologies are a-Si, copper indium gallium selenide (CuIn$_x$Ga$_{1-x}$Se$_2$, CIGS), and cadmium telluride (CdTe). a-Si is the most attractive commercially due to the fact that they can use existing silicon solar cell technology for manufacturing. The known drawbacks of poor absorption and degradation are the roadblocks for this technology. The direct band gap semiconductor materials with high absorption, such as CIGS and CdTe are the effective alternatives and they are famous for their stability and higher efficiency. The CdTe and CIGS PVs are having world record efficiency of 22.1% and 22.3% for small area cells, respectively. Moreover, Si-wafer technology is difficult to achieve PV module production cost below € 1/W (1€ is about US$1.35), which is considered essential for cost-competitive generation of solar electricity (Upadhyaya et al., 2015). This was overcome with the large production capacity of c-Si modules in China which brought the cost below $0.6/W and as low as $0.5/W. While the European market could not cope with this low price and almost collapsed, it was believed the low module cost from China was perhaps an artificial price, and not only due to other factors, such as low-cost of c-Si, cheap labor, etc. (Upadhyaya et al., 2015). This has created turbulent market and global controversies over module pricing which lead to legal action against the Chinese manufacturers (Candelise, 2012). This has a clear message for TF PV manufacturers to improve their efficiency beyond 14% in module level and stay in the PV market.

The interest in TF PV technology started in the early 1980s. The real breakthrough has come during the last decade as more industries have become interested in TF PV technologies. TF PV is more suitable for building integrated applications with the aim to reach 22, 214 MW production by 2020. To make the TF PV alternative to silicon technologies, it has to match or exceed the current c-Si technology. This can be achievable with the careful manufacturing process of deposition of materials with high quality and fine-tuned suitable parameters. However, once optimized, these methods provide over an order of magnitude cheaper processing cost and low energy payback time, which is certainly a big advantage. For example, even at lower module efficiencies as compared to c-Si, the cost of CdTe modules has lower payback time owing to less energy intensive and easier processing steps involved (Upadhyaya et al., 2015).

These alternative materials to c-Si that have more interesting semiconducting materials apart from the technologies mentioned above are: gallium arsenide (GaAs), indium phosphide (InP), zinc phosphide (Zn_3P_2), copper sulfide (Cu_2S), copper indium diselenide (CIS), and copper-zinc tin diselenide (CZTS). Multijunction solar cells based on III-V materials (GaAs, InP, GaSb, GaInAs, GaInP, etc.) show high efficiency, exceeding 35%, but due to the high production cost and low abundance of their constituent materials, these solar cells are not considered suitable for cost effective terrestrial applications, though they are still very important for space PV applications. Table 2.1 shows the important materials and their highest efficiencies achieved so far.

TF technology has an advantage of flexible manufacturing, having rigid and flexible substrates (metallic, flexible, and cloths) which will make them suitable for space applications, building integrated photovoltaics (BIPV), and flexible electronics. The key advantages of the TF technologies are:

TABLE 2.1 Confirmed Terrestrial Cell and Submodule Efficiencies Measured Under the Global AM1.5 Spectrum (1000 W/m²) at 25°C

PV Technology	Efficiency (%)	J_{sc} (mA/cm²)	V_{oc} (V)	FF (%)
Silicon				
Si (crystalline)	25.6±0.5	41.80	0.740	82.7
Si (multicrystalline)	21.25±0.4	39.80	0.667	80.0
Si (thin transfer submodule)	21.2±0.4	38.50	0.687	80.3
Si (thin film minimodule)	10.5±0.3	29.70	0.492	72.1
III-V cells				
GaAs (thin film)	28.8±0.9	29.68	1.122	86.5
GaAs (multicrystalline)	18.4±0.5	23.20	0.994	79.7
InP (crystalline)	22.1±0.7	29.50	0.878	85.4
Thin film chalcogenide				
CIGS (cell)	21.0±0.6	35.70	0.757	77.6
CIGS (minimodule)	18.7±0.6	35.29	0.701	75.6
CdTe (cell)	21.0±0.4	30.25	0.876	79.4
Amorphous/microcrystalline Si				
Si (amorphous)	10.2±0.3	16.36	0.896	69.8
Si (microcrystalline)	11.8±0.3	29.39	0.548	73.1
Dye sensitized				
Dye	11.9±0.4	22.47	0.744	71.2
Dye (minimodule)	10.7±0.4	20.19	0.754	69.9
Dye (submodule)	8.8±0.3	18.42	0.697	68.7
Organic				
Organic thin film	11.0±0.3	19.40	0.793	71.4
Organic (minimodule)	9.7±0.3	16.47	0.806	73.2
Perovskite				
Perovskite thin film	15.6±0.6	19.29	1.074	75.1
Multijunction				
Five junction cell (bonded)	38.8±1.2	9.56	4.767	85.2

Continued

TABLE 2.1 Confirmed Terrestrial Cell and Submodule Efficiencies Measured Under the Global AM1.5 Spectrum (1000 W/m²) at 25°C—cont'd

PV Technology	Efficiency (%)	J_{sc} (mA/cm²)	V_{oc} (V)	FF (%)
InGaP/GaAs/InGaAs	37.9±1.2	14.27	3.065	86.7
GaInP/Si (mech. stack)	29.8±1.5	14.10/22.70	1.46/0.68	87.9/ 76.2
a-Si/nc-Si/nc-Si (thin film)	13.6±0.4	9.92	1.901	72.1
a-Si/nc-Si (thin film cell)	12.7±0.4	13.45	1.342	70.2

FF, fill factor; *J_{SC}*, short circuit current density; *V_{OC}*, open circuit voltage.
(Adapted from Green, M.A., Emery, K., Hishikawa, Y., Warta, W., Dunlop, E.D., 2015. Prog. Photovolt. 23, 805–812 Solar Efficiency table version 46.)

- Much less material usage (1–2 µm of material thickness is sufficient to harness more than 90% of the incident solar light) due to the higher absorption coefficient (~10^5 cm^{-1}) that is about 100 times higher than c-Si.
- TF solar cells have lower energy payback time than c-Si PV. Estimation suggests that CdTe has the lowest payback time among all PV technologies. With recent improved cell and modules efficiencies of 17%, the payback time could be as low as 6 months (Upadhyaya et al., 2015).
- Less absorption losses and enhanced collection due to the better heterojunction formation and device engineering. Easy to have monolithic connections and minimize area losses.
- Easy manufacturing such as roll-to-roll (R2R) process and stacking of devices as tandem and multijunction devices could cover the full solar spectrum which can lead to as high as 67% efficiency theoretically.

In the following sections, the front line TF technologies such as a-Si, CdTe, and CIGS are discussed in detail relating to their material and device aspects, current status, and issues related with environmental concerns.

2.3.1 TF a-Si Solar Cells

Amorphous-Si is one of the most studied materials for TF solar cells. The short range of atomic ordering, less than 1 nm, compared to crystalline Si, which has up to a few centimeter size in the single crystals. a-Si has a disordered lattice showing localized tetrahedral bonding schemes, but with broken Si-Si bonds of random orientation. These broken (or unsaturated) bonds are called "dangling bonds" and contribute to the defect density in the material and will lead to predominant localized defect states. This will be overcome by effective passivation through hydrogen to reduce the dangling bonds density. This will alter the defect distribution and hence, the optical and electronic properties. The absorption coefficient (α), of c-Si (monocrystalline silicon wafers) and microcrystalline TF Si have more or less the same onset of transition, but μc-Si has higher α in low wavelength (λ) region. It plays an important role in material consumption and α for μc-Si is lower than of a-Si, therefore thicker μc-Si layers are required compared to a-Si. The flexible band gap according to the dopant such as O, C, and Ge has a-Si:H band gaps of 2.2–1.1 eV. This is more suitable for the graded type band gap materials for PV applications to maximize the light absorption within the spectrum.

a-Si materials can be prepared through a variety of technologies due to its feature of taking wide temperature from room temperature to 400°C. The room temperature deposition will allow for use of a variety of substrates viz. glass, metal, and plastic. Silane is the most important precursor to be used in the a-Si:H deposition and it is associated with different process, like chemical vapor deposition (CVD), plasma enhanced chemical vapor deposition (PECVD), and glow discharge CVD, etc. However, the deposition temperature for a-Si:H is around 500°C to incorporate hydrogen. The low temperature deposition leads to the predissociation of SiH_4 and room temperature deposited layers give rise to inferior quality and efficiency. This was overcome with PECVD and glow discharge CVD. Alternative deposition methods using the hot wire CVD (HWCVD) technique, electron cyclotron resonance reactor, and also the combination of HWCVD and PECVD are being carried out to increase the deposition rate higher than 10 Å/s to achieve industrial scale deposition.

The conventional p-n junction configuration for a-Si:H based solar cells is not very suitable due to their large number of defect states. So it needs a solar cell configuration of p-i-n junction where an intrinsic layer of a-Si:H is sandwiched between the n- and p-type doped layers of a-Si:H or its alloys. Even though the initial results were very encouraging in the 1970s, an inherent problem of light dependent degradation on their performance under continuous light exposure had created constrains in its market. The light induced degradation will be tackled using different approaches, such as thickness of intrinsic layer of ∼300 nm and tandem cells using double and triple junctions. Fig. 2.2 shows the multijunction solar cells with different architecture.

The hybrid type of devices with a-Si cells with c-Si and other semiconductor materials are also developed. Another significant development in the design is the formation of a thick/thin type of interface structure (hetero-structure) between a-Si:H layer and c-Si wafer, referred to as HIT (heterojunction with intrinsic TF layer) cells are shown in Fig. 2.3 and achieved the efficiency close to 21% over a cell area 101 cm^2.

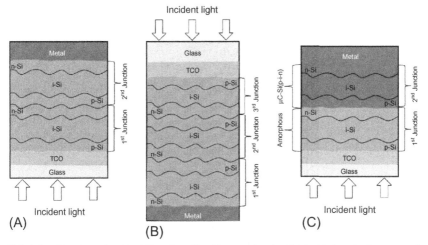

FIG. 2.2 The schematics of multijunction cell architecture showing (A) double junction "superstrate" configuration, (B) triple junction in "substrate" configuration, and (C) "micromorph" junction in "superstrate" configuration.

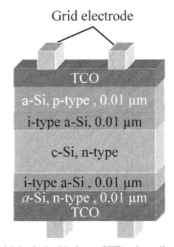

FIG. 2.3 Heterojunction with intrinsic thin layer (HIT) solar cell architecture.

2.3.2 Cadmium Telluride Solar Cells

Cadmium telluride (CdTe) is a II-VI metal chalcogenide with direct band gap of ~1.45 eV, very high optical absorption (10^5 cm^{-1}), and p-type conductivity, making it an ideal material for PV application. The CdTe TF can be deposited using a number of processes and needs good stoichiometric at over 400°C. The usual architecture (Fig. 2.4) of the CdTe solar cells is comprised of a p-type CdTe absorber layer and n-type CdS based window layer forming

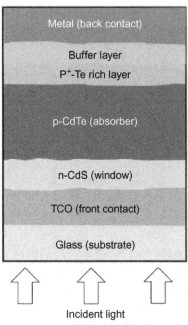

Incident light

FIG. 2.4 Schematic presentation of CdTe/CdS solar cell in "superstrate" configuration showing different layers and their nomenclature.

a heterojunction. It has been started around 1962 with 5–6% efficiency and now it has been reported 22.1% recently.

The CdTe is most attractive for its chemical simplicity and the robust stability. This accounts for the excellent material stability and device performance. It is proved that CdTe is not only stable for terrestrial applications, but it has also been demonstrated that CdTe has excellent stability under high energy-photon and electron irradiation for space applications which is superior to Si, GaAs, CIGS, etc.

CdTe TF solar cells are grown on rigid and flexible substrates. Both substrate and superstrate configurations have achieved conversion efficiencies 14% and 22%, respectively. The superstrate configuration is sequentially deposited on glass substrates as transparent conducting oxide (TCO), CdS, CdTe, and the metal back contact to finish the device structure. TCO layers are having prerequisite with an electron affinity below 4.5 eV to form an ohmic contact. Cadmium sulfide (CdS) is a most common n-type window layer with energy band gap of 2.4 eV. A typical thickness of CdS layer used in solar cells is in the range of 10–500 nm. The absorber layer is made from direct band gap CdTe TFs. Due to its advantage over the light absorption, it is sufficient to have only 2 μm thick material. It can be grown through a variety of vacuum and non-vacuum methods. The most commonly used high temperature methods are closed space sublimation (CSS), vapor transport (VT), or vapor transport deposition (VTD) with deposition temperature above 500°C. Very famous low-cost methods such as electrodeposition (ED), screen printing, and chemical spraying (CS) are done around 450°C and classified as low temperature processes. The poor PV property of the as-deposited CdTe/CdS solar cells needs annealing treatment under Cl-O ambient between 350°C and 600°C. This is known as "CdCl$_2$ treatment" or "junction activation" treatment and efficiency increases by a factor of 3–5.

2.3.3 Cu (In Ga)Se$_2$ Solar Cells

CIGS is one of the compound semiconductors made from the I-III-VI series of periodic table. The elements such as CIS, copper gallium diselenide and their mixed alloys CIGS are often prepared in a wide range of compositions for many different applications. The solar cells out of these materials can be achieved by changing the stoichiometry and extrinsic doping. Change in the [Ga]/[In+Ga] ratio, the band gap of CIGS can be varied continuously between 1.04 and 1.68 eV. The efficiency of the solar cell devices is based on the band gap ranging from 1.15 to 1.25 eV of CIGS. Other chalcopyrites compounds such as CuInS$_2$ and CuInTe$_2$ were also equally investigated. However, the efficiency of the CIGS solar cells is higher compared to the other chalcopyrites. The CIS solar cells were developed with single-crystal material and reported higher conversion efficiencies, and Siemens (Shell Solar) has started first industrial production based on the Arco solar technology (Upadhyaya et al., 2015). Many companies were involved in the production of CIS manufacturing through different low-cost nonvacuum deposition methods.

CIGS solar cells have a similar type of architecture as CdTe TF solar cells as shown in Fig. 2.5. CIGS solar cells can be grown on glass as well as metal and

Incident light

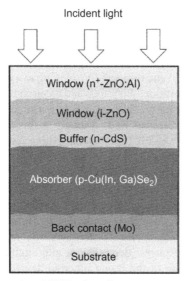

FIG. 2.5   Schematic presentation of CIGS solar cell.

polymer foils. Molybdenum (Mo) is the most commonly used electrical back contact material for CIGS solar cells. CIGS layer is an absorber layer and can be grown with a variety of deposition methods such as co-evaporation process, three stage process, etc. The important role in these methods is the selenization of the precursor materials to achieve CIGS. There is a substantial interest in using nonvacuum methods for CIGS deposition such as ED from a chemical bath, ED followed by an RTP selenization treatment, etc.

The n-type conductor on CIGS absorber layer is usually a CdS layer with band gaps between 2.0 and 3.4 eV to form a heterojunction in CIGS solar cells. CdS for highest-efficiency CIGS cells is commonly grown by chemical bath deposition (CBD), which is a low-cost, large-area process. The front contact for CIGS is mostly used from a TCO layer with band gaps of above 3 eV.

2.4   THIRD GENERATION PV TECHNOLOGIES

2.4.1   Dye Sensitized Solar Cell

Currently, different types of solar technologies are classified into third generation solar cells. The main technologies are DSSCs, organic solar cells (OPV), and perovskite solar cells (PSC). The foremost aim of these technologies is to innovate the PV manufacturing into low-cost electricity supply. DSSCs are the most studied PV technology in the last couple of decades with a large number of industries and academic research involved to improve the efficiency. More than 2000 international patents have been filed for this technology which has started with dye sensitized and later, organic solar cells. DSSCs are following a completely different device approach of bulk-heterojunction (BHJ) formation in the nanometer length regime. It consists of nano-composites of mesoporous TiO_2 and inorganic dyes (typically ruthenium complexes). So far it has reached around 13% efficiency through the use of rather inexpensive and abundant materials (O'Regan, 1991).

DSSCs are working in principle by excitonic electron-hole pair generation upon illumination of light and dissociation, leading to a charge separation at the nano-interface. The absorption of photons and electron transport is mediated via two dissimilar constituents viz. dyes and n-type nanocrystalline TiO_2, respectively. The holes are transported through a hole-transporting material (HTM), which could be a redox liquid electrolyte or ion conducting polymeric electrolyte (quasi-solid state) or quantum dot sensitized or hole conducting conjugated polymer material or small molecules (solid state). This is very different from the conventional p-n junction solar cells. The efficiency of the lab scale devices is remarkable, the liquid junction appears to have now reached a state of stagnation in efficiency which is hindering the technology to launch effectively into PV market. The issues like the lack of sensitizer dyes with wider spectral coverage, atmospheric degradation, and engineering issues, such as encapsulation and sealing of the liquid junction device, which had prevented this technology from making a mark in commercialization. Global DSSC market value was estimated to be USD 49.6 million in 2014 and expected to grow over 12% from 2015 to 2022. This is due to the environmental awareness among the public over the years and associated impacts of fossil fuel-based power plants. The anticipated market for this technology within BIPV applications, and building applied photovoltaics (BAPV), and net growth would exceed USD 30 million by 2022. The road map to achieve zero emission buildings is to integrate renewable energy resources; DSSC is one of the best options due to its ability to work in diffused light conditions.

2.4.2   Organic Solar Cell

A typical OPV device consists of one or several photoactive materials sandwiched between two electrodes. In a bilayer device, light is absorbed in the photoactive layers composed of donor and acceptor semiconducting organic materials to generate photocurrents. The donor material donates electrons and mainly transports holes and the acceptor material withdraws electrons and mainly transports electrons. Due to the concentration gradient, the excitons diffuse to the donor/acceptor interface (exciton diffusion) and separate into free holes (positive charge carriers) and electrons (negative charge carriers) (charge separation). A PV is generated when the holes and electrons move to the corresponding electrodes by following either donor or acceptor phase (charge extraction). A primary advantage of OPV technology over inorganic counterparts is its ability to be utilized in large areas and flexible solar modules, specially facilitating R2R production. Additionally, manufacturing cost can be reduced for organic solar cells due to their lower cost compared to silicon based materials and the ease of device manufacturing. However, to catch up with the performance of silicon based solar cells, both donor and acceptor materials in an OPV need to have good extinction coefficients, high stabilities, and good film morphologies. Since the donor plays a critical role as the absorber to solar photon flux, donor materials require wide optical absorption to match the solar spectrum. Another basic requirement for ideal donor/acceptor is a large hole/electron mobility to maximize charge transport. The significant improvement of OPV device performance has been accomplished by introducing various OPV architectures, such as BHJ and inverted device structures, and developing low band gap conjugated polymers and innovative organic small molecules as donor materials. It is estimated that the OPV market will rise to USD 87 million by 2023 (Luna-Rubio et al., 2012).

2.4.3   Perovskite Solar Cell

PSC have emerged from solid state DSSCs. However, the efficiency of the DSSCs are very much restricted and perovskite solid state absorbers have the ability to increase the efficiency over 20%. It has emerged as a new type of

PSC and reached as high as 22.1% photon conversion efficiency (PCE). The PSC realized the breakthrough efficiency in late 2012 with the introduction of solid state hole-transporting layers within the solar cell, which resulted in stable efficiencies close to 10% (Sahu, 2015). In a short span of 5 years, efficiency has reached 22.1%, which could be an energy revolution in the future. Generally, PSC have TCO/metal oxide/perovskite/HTM/metal architecture. Hybrid organic-inorganic lead halide perovskites can combine functions of light absorption, n-type conduction, and p-type conduction, the perovskite absorbs light and electron-hole pairs are created in the material, which can possibly evolve towards the formation of excitons after thermalization of the carriers. Charge separation can then occur through two possible primary reactions: injection of photogenerated electrons into metal oxide nanoparticles, and/or injection of holes into a HTM, such as spiro-OMeTAD. PSC can be fabricated by low-cost solution processed methods both on glass and flexible substrates. Moreover, global PSC market value is estimated to be USD 214 million by 2025 (Suganthi and Samuel, 2012).

2.4.4   Concentrating PVs

Solar cells are usually very expensive. An effective way of reducing the cost of PV systems is either by reducing solar cell manufacturing cost or illuminating solar cells with a higher light intensity than is available naturally. In the latter case, solar cells convert the additional power incident without significant loss of efficiency. Either trapping light within screen-printed solar cells, or using reflective/refractive devices to increase the luminous power flux on to the solar cell surface can do this. Static parabolic trough concentrators for different receiver locations have been reported for PV applications. A one-axis tracking parabolic trough mirror and a three-dimensional second stage compound parabolic PV concentrator with a geometrical concentration ratio of 200 has achieved an electrical efficiency of 26%. There are several lens and concentrator ratios that have tested so far and achieved an efficiency on multijunction solar cells up to 44%.

2.5   CONCLUSIONS

Solar energy systems can reduce CO_2 emission levels and lower the electricity price. The PV market is booming with a phenomenal surging growth rate of over 40% for more than a decade, despite a downturn observed in the worldwide economy recently. The cumulative PV production capacity was $\sim$140 GW by 2013. An upsurge in the PV production and an artificial module price crisis created by Chinese manufacturers has brought a radical drop in the module prices well below $0.6/Wp since 2011. Although this strategy has wiped away the PV industries in Europe, it has caused a PV manufacturing base shift towards Asia with China, Taiwan, and Korea taking this opportunity and leading the show through their natural advantage of a cheaper material supply and cheaper human resources available. TF PV has clearly demonstrated an excellent potential for cost effective generation of solar electricity using CdTe technology by First Solar in the United States and CIGS technology by Solar Frontier in Japan, currently aiming to produce 2 GW capacities annually. The related environmental concern has to properly be taken care of by the industries to avoid the fatal accidents, or worse environmental and health issues. Further development of research could lead to better materials with less effect. Chapter 3 discusses the various processes involved in making solar cells and their concerns/impacts.

ABBREVIATIONS

a-Si	amorphous silicon
ARC	antireflection coating
BAPV	building applied photovoltaics
BHJ	bulk-heterojunction
BIPV	building integrated photovoltaics
c-Si	crystalline silicon
CBD	chemical bath deposition
CdS	cadmium sulfide
CdTe	cadmium telluride
CIGS	copper indium gallium diselenide
CS	chemical spraying
CSS	closed space sublimation
CVD	chemical vapor deposition
Cu_2S	copper sulfide
CZTS	copper-zinc tin diselenide
DSSCs	dye sensitized solar cells
ED	electrodeposition
EIA	Energy Information Administration
GaAs	gallium arsenide
GaInAs	gallium indium arsenide
GaInP	gallium indium phosphide
GaSb	gallium antimonite
GHG	greenhouse gas
HTM	hole transporting material
HWCVD	hot wire chemical vapor deposition
IAEA	International Atomic Energy Agency
InGaAs	indium gallium arsenide
InP	indium phosphide
KMT	kilometric tons
LC	life cycle
Mo	molybdenum
NREL	National Renewable Energy Laboratory
OPV	organic photovoltaics
PCE	photon conversion efficiency
PECVD	plasma enhanced chemical vapor deposition
PV	photovoltaic

R2R	roll-to-roll
Si	silicon
Spiro-OMeTAD	2,2′,7,7′-tetrakis-(*N*,*N*-di-4-methoxyphenylamino)-9,9′-spirobifluorene
TCO	transparent conducting oxide
TF	thin film
VTD	vapor transport deposition
WEC	World Energy Council
Zn$_3$P$_2$	zinc phosphide

REFERENCES

Barthelmie, R., Pryor, S., 2014. Nat. Clim. Chang. 4, 684–688.

Braga, S.P.M.A.F.B., Zampieri, P.R., Bacchin, J.M.G., Mei, P.R., 2008. Sol. Energy Mater. Sol. Cells 92, 418–424.

Candelise, C., 2012. UKERC technology and policy assessment, cost methodologies project: PV case study. http://www.ukerc.ac.uk/publications/cost-methodologies-project-pv-case-study. html.

Chopra, K.L., Paulson, P.D., Dutta, V., 2004. Prog. Photovolt. Res. Appl. 12, 69–92.

Fally, J., Fabre, E., Chabot, B., 1986. Polyx photovoltaic technology: progress and prospects. Revue de l'Energie 37 (385), 761–768.

Green, M.A., Emery, K., Hishikawa, Y., Warta, W., Dunlop, E.D., 2015. Prog. Photovolt. Res. Appl. 24, 3–11.

http://www.greentechmedia.com/research/report/polysilicon-2015-2018, 2014.

Jäger-Waldau, A., 2004. Sol. Energy 77, 667–678.

Luna-Rubio, R., Trejo-Perea, M., Vargas-Vázquez, D., Ríos-Moreno, G., 2012. Sol. Energy 86, 1077–1088.

Masson, G., Latour, M., Biancardi, D., 2012. Global Market Outlook for Photovoltaics Until 2016. European Photovoltaic Industry Association Brussels. http://www.epia.org/fileadmin/user_ upload/Publications/Global-Market-Outlook-2016.pdf (accessed 25.12.15.).

Masson, G., Latour, M., Rekinger, M., Theologitis, I.-T., Papoutsi, M., 2013. Global Market Outlook for Photovoltaics 2013-2017. European Photovoltaic Industry Association. http:// www.epia.org/fileadmin/user_upload/Publications/GMO_2013_-_Final_PDF.pdf (accessed 25.12.15.).

Müller-Fürstenberger, G., Wagner, M., 2007. Ecol. Econ. 62, 648–660.

http://www.nrel.gov/docs/fy13osti/56487.pdf (accessed 20.12.15.).

http://www.nrel.gov/ncpv/images/efficiency_chart.jpg (accessed 23.12.15.).

O'Regan, M.G.B., 1991. Nature 353, 737–739.

Power and Energy Industry of HIS Technology, 2015. Top Solar Power Industry Trends for 2015. https://www.ihs.com/pdf/Top-Solar-Power-Industry-Trends-for-2015_2139631109155583632. pdf.

Sahu, B.K., 2015. Renew. Sustain. Energy Rev. 43, 621–634.

Suganthi, L., Samuel, A.A., 2012. Renew. Sustain. Energy Rev. 16, 1223–1240.

Upadhyaya, H.M., Sundaram, S., Ivaturi, A., Buecheler, S., Tiwari, A.N., 2015. Energy Efiiciency and Renewable Energy Hand Book. CRC Press, Boca Raton, FL, pp. 1423–1474 (Chapter 45).

Weisser, D., 2007. Energy 32, 1543–1559.

Chapter 3

Potential Environmental Impacts From Solar Energy Technologies

3.1 INTRODUCTION

Electricity produced from photovoltaic (PV) solar panels is clean and carbon-free compared to fossil fuels based electricity production. PV panels are safe and don't produce toxic air or greenhouse gas (GHG) emissions. However, taking PV as a product from raw materials to solar cells production involves many potential environmental, health, and safety hazard materials. The full product lifecycle analysis in recent years has raised more concerns about hazardous materials and their recycling/disposal, even after 25 years. Most of the solar cell technologies are using hazardous chemicals to achieve higher conversion efficiency. The cleanliness of the materials often decides the conversion efficiency of the solar cells and panels. The most efficient solar cells are from single crystalline solar cells, which need tremendous effort and toxic chemical usage. The processes involved in making silicon chips have many hazardous materials. The extraction of raw materials for silicon-based PV; thin-film (TF) PV (CdTe, CIGS, InGaAs, etc.) and nanostructured solar cells can also pave way to potential environmental, health, and safety. At the same time, emerging solar PV production technologies such as organic solar cells and perovskite solar cells are also involved with unknown health and environmental hazards.

Regardless of the specific PV technologies, PV can generate 89% of potential harmful air emission per kilowatt-hour than conventional fossil fuel (Fthenakis et al., 2008; Fthenakis and Alsema, 2006; Mason et al., 2006). The production of PV cells involves chemical hazards related to the materials' toxicity, corrosivity, flammability, and explosiveness. This chapter discusses the materials processing in major technologies such as silicon, TF, and new generation solar cells, particularly about the toxicity of the materials to humans and biota. In this context, this chapter

1. discusses different solar cell technologies and their manufacturing processes,
2. summarizes the potential health, safety, and environmental hazards associated with different PV technologies, and

Solar Photovoltaic Technology Production. http://dx.doi.org/10.1016/B978-0-12-802953-4.00003-2
23

3. acceptable limit of exposure of chemicals and hazardous air emission.

Apart from the hazardous materials used during the process, the PV power stations have potential impact on land use, habitat loss, and water use (for solar thermal power plants), and life cycle global warming emissions. The solar power plants have less opportunity to use the land for agriculture which leads to land degradation and habitat loss.

3.2 PROCESSING AND LIFE CYCLE OF PV SYSTEMS

The simplified overview process diagram (Fig. 3.1) illustrates the basic key steps involved in the PV lifecycle. The lifecycle of PV is a framework for considering the environmental inputs and outputs of a product or process from cradle to grave (Fthenakis and Kim, 2011; Raugei et al., 2007; Stoppato, 2008). The industrial scale PV production depends on the following major areas in product design and development:

1. *Raw materials mining and refining toward the necessary requirements of PV materials feed stock:* The primary hazards are exposure to and inhalation of kerf dust.

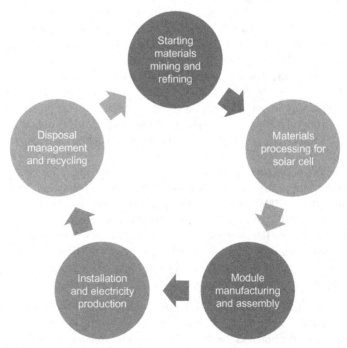

FIG. 3.1 Overview of photovoltaic solar cell materials process and life cycle.

2. *Materials processing:* PV technologies use many hazardous materials during the materials process which can be classified as potential environmental and health issues of exposure to sawing the silicon ingots into wafers, and exposure to solvents, such as nitric acid, sodium hydroxide, and hydrofluoric acid, used in wafer etching and cleaning, as well as reactor cleaning.
3. *Product design and development:* This step involves solar cell soldering, arranged into module and complete assembly with several other layers, such as back plate, sealant materials, toughened high transmission glass, electrical connections and framing of the module. These processes involve occupational health hazards.
4. *Installation and electricity production:* Readily available products will be installed either in building roofs, or in power plants.

The lifecycle analyses of the solar modules include the recycling of the modules after its lifetime. There have been plenty of lifecycle analyses published so far. The Life Cycle Assessment (LCA) Harmonization Project led by the National Renewable Energy Laboratory (NREL) suggested harmonious ways to analyze the lifecycle GHG emission for PV systems. So far, many studies have reported on crystalline silicon (c-Si) (monocrystalline and multicrystalline) and TF [amorphous silicon (a-Si), cadmium telluride (CdTe), and copper-indium gallium diselenide (CIGS)] which were analyzed by NREL. Fig. 3.2 is a comparison of the lifecycle stages for PV system and coal based power plants.

LCA of PV power plants shows that the majority of the GHG emission is during the materials extraction and preparation and module manufacturing.

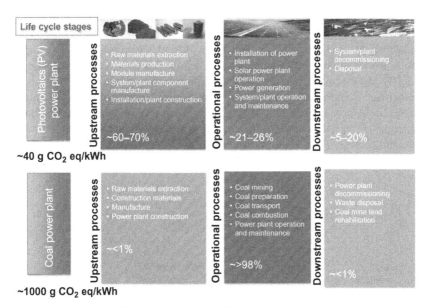

FIG. 3.2   Comparison of the lifecycle stages for PV system and coal based power plant.

The lifecycle GHG emissions for PV power systems are compared with other electricity generation technologies by NREL harmonization which shows that the total lifecycle GHG emissions from solar PV systems are similar to other renewables and nuclear energy, and much lower than coal (http://www.nrel.gov/analysis/sustain_lcah.html). However, the materials manufacturing and module preparation processes can leave worse effects on the environment which is discussed in the following section.

3.3   MANUFACTURING OF SILICON

Silicon-based solar cell production started from the microelectronics industry. The knowledge of making silicon wafers was majorly adopted from the microelectronics industry. Different methods have been adopted to manufacture silicon. The quality/purity of the silicon will decide the process. According to the available purity, we can classify the silicon into metallurgical grade and semiconductor grade. The metallurgical grade silicon (MG-Si) can be approximately pure up to 99.6% and semiconductor devices need much purer. The purity of the silicon is achieved through a series of chemical processes. Commonly, three basic steps are involved in silicon production, such as reduction of silica, cooling/crushing, and packaging. The first and foremost step in conversion of sand into silicon is reduction. Section 3.3.1 will discuss the different silica reduction methods available.

3.3.1   Reduction of Silica

Silica (SiO_2) is a starting material for silicon, which is sand or quartz. The removal of oxygen or reducing silica to silicon is a key processing step in silicon production. Fig. 3.3 shows the different methods used to reduce the silica to silicon. MG-Si is the starting material to produce pure silicon for PV and electronics applications. Silicon is produced industrially by carbothermic reduction of silica in submerged-arc electric furnaces. The other processes are seldom used in the silicon manufacturing on an industrial scale. Here we consider only the carbothermic reduction process. The process can be written simply as:

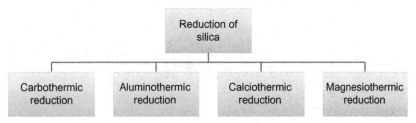

FIG. 3.3   Different methods of silica reduction.

$$SiO_2 + 2C = Si + 2CO(g) \qquad (3.1)$$

The carbothermic process includes silicon source and reduction agents of coke, coal charcoal, and wood chips. This method will lead to 99.0 wt.% of Si and other impurities, such as Fe, Al, Ti, Mn, C, Ca, Mg, B, P, and so on. The impurities will be removed through further chemical processes to result in high pure silicon for solar and electronic applications.

Reduction of silica by carbon is highly endothermic and needs a very high processing temperature. The reduction of silica using carbon in a typical industrial furnace (Schei et al., 1998) has a shell diameter of about 10 m as shown in Fig. 3.4. The carbon electrodes submerged into the charge material and heats up to about 2000°C to form molten silicon. Molten silicon is a reduced form of silicon dioxide. The oxygen removal process has typically been done through high temperature (2000°C) process to reduce silica to molten silicon. This process is usually done through a slag treatment or gas purging. This will lead to pure silicon without having oxide and carbide particles. After refining, the

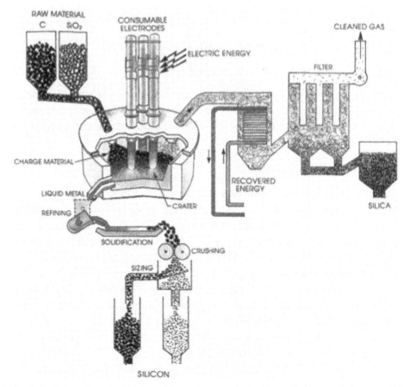

FIG. 3.4 Schematic of silicon metal production plant. *(From Salazar-Villalpando, M.D., Neelameggham, N.R., Post Guillen, D., Pati, S., Krumdick, G.K. (Eds.), 2012. Energy Technology 2012: Carbon Dioxide Management and Other Technologies. John Wiley and Sons.)*

molten alloy is allowed to cool in a suitable mold to attain a specific size. During the refining process, the gas used for purging has toxic elements after its filtrations, mainly sulfur dioxide.

3.3.2   Purification of Silicon

The purification of the silicon is done primarily through a chemical approach. The highest purity of the silicon is achieved through reduction and pyrolysis of volatile compounds such as $SiHCl_3$, SiH_2Cl_2, SiH_4, $SiCl_4$, and $SiHBr_3$. This is famously known as chloride-hydride technology. The purification methods using gasification of MG-Si and deposition of silicon was established in the 1950s and known as the Siemens process (Gribov and Zinov'ev, 2003; Tsuo et al., 1998). The typical chemical processing is given in the reactions (3.2–3.5). According to reaction (3.2), the MG-Si is reacting with HCl to produce trichlorosilane ($HSiCl_3$) at 500°C and 30 MPa temperature and pressure, respectively.

$$5(MG)Si + 16HCl = 4HSiCl_3 + 6H_2 + SiCl_4 \qquad (3.2)$$

$$2SiHCl_3 = SiH_2Cl_2 + SiCl_4 \qquad (3.3)$$

$$2SiH_2Cl_2 = SiHCl_3 + SiH_3Cl \qquad (3.4)$$

$$2SiH_3Cl = SiH_2Cl_2 + SiH_4 \qquad (3.5)$$

The reactions (3.3–3.5) are responsible for the redistribution of chlorine and hydrogen atom from the metal chloride compounds and generating silane gas. The steps in the process are necessary to distill diboride and other impurities.

3.3.3   Crystalline Silicon (c-Si)

The starting material for production of c-Si wafers and silicon chips is silica (SiO_2). However, silicon for semiconductor use must be much purer. Intense chemical processing is involved to make the MG-Si into semiconductor grade silicon. Usually, MG-Si is exposed to hydrochloric acid and copper to produce a gaseous trichlorosilane ($HSiCl_3$). The gaseous trichlorosilane is then distilled to remove the chlorinated metals as prevailing impurities remaining in the process. The process is followed by a reduction with hydrogen to produce silane (SiH_4) gas. The silane gas is heated further to grow monocrystalline silicon crystals. The monocrystalline silicon rods are crushed after taken from the furnace to introduce them into the reactor at high temperature and high pressure. The silane gas is being used to deposit additional silicon on to the rods to grow the desired diameter. For the multicrystalline silicon, it is poured into crucibles and cooled into blocks or ingots. The purity of the silicon crystals is extremely pure (from 99.99999% to 99.9999999%) for microchips and other electronic industry usage. The whole process involves a high-energy intensive and expensive process with large amount of waste production. It is estimated that 80% of the initial MG-Si is lost in the process (Goetzberger and Hoffmann, 2005; McEvoy et al., 2003). The purity of the silicon decides the ways to refine and methods to purify. Compared to single crystalline silicon process, other silicon grades, such as multicrystalline, or amorphous silicon, adopt less chemical environment.

3.3.4   Hazardous Materials Used in Si Processing and Manufacturing

The silica is mined to use in different purposes. Most of the silica is used in the steel industry and a small fraction is going to the semiconductor and PV industries. The main hazardous material in the mining and c-Si wafer production is dust. The c-Si waste dust is called kerf and up to 50% is lost in air and water used to rinse wafers (McEvoy et al., 2003). The mining of metallurgical grade silica can produce silica dust, which can lead to severe lung diseases. The personal productive equipment like respiratory masks can solve the health issues to a certain level of exposure. However, the silicon particulate will pose inhalation problems for the workers. Table 3.1 lists the materials used in the silicon and related health issues caused.

TABLE 3.1 Hazardous Materials and Associated Health Issues in Silicon Processing

Hazardous Material	Environment Used	Health Issues
Metallurgical grade silica/ silica dust	Mining	Silicosis, a severe lung disease
Hydrochloric acid	Reduction chamber	Irritant/corrosive
Carbon tetrachloride	Etchant	Liver cancer, greenhouse gas
Chlorosilanes	a-Si and x-Si deposition	Irritant
Diborane	a-Si dopant	CNS, pulmonary
Germane	a-Si dopant	Blood, CNS, kidney
Hydrogen	a-Si deposition	Fire hazard
Hydrogen fluoride	Etchant	Irritant, burns, bone, teeth
Lead	Soldering	CNS, GI, blood, kidney, reproductive
Nitric acid	Wafer cleaning	Irritant, corrosive

Continued

TABLE 3.1 Hazardous Materials and Associated Health Issues in Silicon Processing—cont'd

Hazardous Material	Environment Used	Health Issues
Phosphine	a-Si dopant	Irritant, CNS, GI, flammable
Phosphorous oxychloride	x-Si dopant	Irritant, kidney
Selenium compounds	CIS deposition	Irritant
Sodium hydroxide	Wafer cleaning	Irritant
Silane	a-Si deposition	Irritant, fire, and explosion hazard
Silicon tetrafluoride	a-Si deposition	Corrosive, irritant to eye, lung, and skin
Sulfur hexafluoride	Clean the reactors	Greenhouse gas
Sulfur dioxide	Within chamber	Acid rain
Sodium hydroxide/ potassium hydroxide	Remove the sawing damage	Eye, lung, and skin damage
Phosphine/arsine	Doping with Si	Accidental risk/ occupational risk

Although no public health issues were identified with silicon technologies, the environmental issues are identified. The environmental issues are related to the generation of liquid and solid wastes during wafer slicing and etching, and processing and assembling of solar cells (Fthenakis and Moskovitz, 1990). The health-related issues during the processing and chemical handling are massive in this technology. The silane gas is the most hazardous in the production and extremely explosive. The semiconductor industries have reported several accidental releases of silane over the years (Fthenakis, n.d.; Cha, 2008). The silicon tetra chloride ($SiCl_4$) is extremely toxic when it reacts with water and causes skin burns and irritations. There are no environmental regulations available for this extreme environmental hazardous material (Alsema et al., 1996). There are strong regulations in the use of sulfur hexafluoride (SF_6) from the Intergovernmental Panel of Climate Change (IPCC). It is rated one of the most potent GHGs per molecule; as one ton is equivalent to 25,000 tons of CO_2 (Moskowitz and Fthenakis, 1991a). It can create acid rain when reacting with silicon to generate silicon tetrafluoride (SiF_4) and sulfur difluoride (SF_2), or be reduced to tetrafluorosilane (SiF_4) and sulfur dioxide (SO_2). However, other chemicals listed in Table 3.1 are used in the production of c-Si which requires special handling and disposal procedures. The production of monocrystalline silicon is mainly for the electronic industry. However, the growing PV market is surpassing the electronic industry. The additional chemicals, such as ammonium fluoride, nitrogen, oxygen, phosphorous, phosphorous oxychloride, and tin are used in the production line (Fthenakis, 1998). Apart from crystalline and monocrystalline silicon, multicrystalline silicon also requires special chemicals with special handling and operating procedures, including ammonia, copper catalyst, diborane, ethyl acetate, ethyl vinyl acetate, hydrogen, hydrogen peroxide, ion amine catalyst, nitrogen, silicon trioxide, stannic chloride, tantalum pentoxide, titanium, and titanium dioxide (Fthenakis, 1998).

The amorphous silicon (a-Si) materials are often used in TF solar cells and PV panels. The base or substrate materials are metal, glass, plastic, and thin wafers. The a-Si uses silane or chlorosilane gas heated and mixed with hydrogen. As discussed earlier, silane is one of the most hazardous gases and explosives. Several accidents involving silane gas explosions have been reported in the past (Moskowitz, 1995). Another explosive gas used in the process is hydrogen and usually mixed with methane to recycle the hydrogen in waste streams. The methane is also a potential GHG and a potential threat to the environment if it is released. The dopants used in the a-Si such as germane gas are considered to be toxic and damage blood and kidneys (Fthenakis, 2003; Avrutin et al., 2011). The other dangerous chemicals such as hydrochloric acid, hydrofluoric acid, phosphoric acid, sodium hydroxide, acetone, aluminum, chlorosilanes, diborane, phosphine, isopropanol, nitrogen, silicon tetrafluoride, tin, germanium, and germanium tetrafluoride used in the process need special handling to avoid occupational injury (Fthenakis, 1998). Table 3.2 shows the materials used in the silicon PV industry and their potential environmental hazards.

TABLE 3.2 Materials Used in Silicon Photovoltaic Industry and Their Potential Environmental Hazards With Regulatory Measures

Materials	Environmental Hazard	Regulations
Silicon tetrachloride ($SiCl_4$)	React with water and cause environmental hazard	Used in polysilicon manufacturing industries and very little recycling facilities available
Sulfur hexafluoride (SF_6)	Extremely potential greenhouse gas	IPCC regulation
SF_6 can react with silicon to form silicon tetrafluoride (SiF_4) and sulfur difluoride (SF_2), or be reduced to tetrafluorosilane (SiF_4) and sulfur dioxide (SO_2)	Sulfur dioxide (SO_2) is a potential hazard to make acid rain	No particular regulation of sulfur dioxide (SO_2) usage in PV industry as it is a co-created gas during the SF_6 exposure

3.4   TF PV MATERIALS

TF solar cells have been an alternative to c-Si solar cells for over four decades. The TF materials have low absorption coefficient ($\sim 10^3$ cm^{-1}) and narrow band gap ($E_g \sim 1.1$ eV). The conversion efficiency of the TF solar cells with new generation of materials has reached greater than 20%. The most famous semiconducting materials with higher conversion efficiency and available in the market are cadmium telluride (CdTe), gallium arsenide (GaAs), indium phosphide (InP), copper-indium gallium diselenide (CIGS), and copper-zinc tin diselenide (CZTS). Apart from these materials, there are a lot of other materials with electronic and optical properties suitable for solar cell. The TF solar cells have an advantage of multiple junctions to cover the entire solar spectrum with suitable band gap engineering. The multijunction solar cells are based on III-V materials (GaAs, InP, GaSb, GaInAs, GaInP, etc.) with conversion efficiency over 35%. Due to the high production cost and low abundance of their constituent materials, these solar cells are not considered suitable for terrestrial applications, though they are still very important for space PV applications. Table 3.3 shows the hazardous materials used in TF deposition and their related health issues.

TABLE 3.3 Hazardous Materials and Associated Health Issues in Thin-Film Processing

Hazardous Material	Environment Used	Health Issues
Arsine	GaAs CVD	Blood, kidney
Arsenic compounds	GaAs	Cancer, lung
Cadmium compounds	CdTe and CdS deposition CdCl$_2$ treatment	Cancer, kidney
Hydrogen selenide	CIS sputtering	Irritant, GI, flammable
Hydrogen sulfide	CIS sputtering	Irritant, CNS, flammable
Indium compounds	CIS deposition	Pulmonary, bone, GI
Selenium compounds	CIS deposition	Irritant
Tellurium compounds	CIS deposition	CNS, cyanosis, liver

3.4.1   Cadmium Telluride (CdTe) Solar Cells

CdTe is one of the suitable materials with an ideal band gap of ~1.5 eV with high optical absorption quality and high chemical stability. Apart from its suitable materials characteristics, TF CdTe is regarded as one of the leading materials to make cost-effective PV and can offer the price for watt-peak (Wp) below $1 ($0.85) (Fang et al., 2011). The theoretical efficiency of CdTe thin-film solar cells is expected to be 28–30% (Bosio et al., 2006; Fthenakis et al., 1999). CdTe solar panels use the layers of CdTe and cadmium sulfide (CdS). These layers have been made through different chemical and physical deposition methods, such as radio frequency sputtering (R.F. sputtering), close-spaced sublimation (CSS), chemical bath deposition (CBD), electro-deposition (ED), and screen-printing. Cadmium (Cd) is a toxic element which is a by-product of zinc mining and batteries. The scarcity element tellurium (Te) is a by-product of copper, lead, and gold mining. The bottleneck for CdTe technology production is to recycle the Te. The main concern of the cadmium in CdTe is its toxicity which can cause lung carcinogen and has long-term detrimental effects on kidney and bone. However, CdTe as a compound is more stable and less soluble. There are less comparative studies reporting on TF materials and its toxicological data (Fthenakis, 2004). The cadmium emission sources from other than PV are holding much higher values, as shown in Fig. 3.5.

The starting materials to make CdTe TFs are usually a solution of cadmium sulfate ($CdSO_4$) or cadmium chloride ($CdCl_2$), mixed with tellurium dioxide

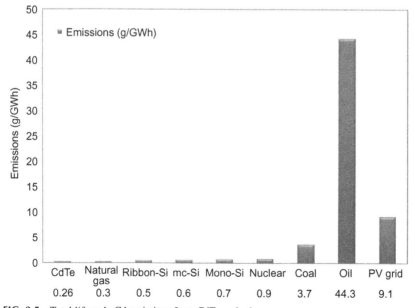

FIG. 3.5   Total lifecycle Cd emissions from CdTe and other sources.

(TeO$_2$) when they are coated through chemical methods. In this process, cadmium-containing water is a potential environmental pollutant. Using careful recycling methods can avoid the cadmium contamination (Moskowitz and Fthenakis, 1991a). Depositing CdS is also important to get higher conversion efficiency. Most of the methods reported to deposit CdS are using a mixture of cadmium sulfate (CdSO$_4$), thiourea (CS(NH$_2$)$_2$), and ammonia (NH$_3$). The waste from these methods is very minimal and disposed as solid waste (S.V.T. Coalition, 2009). The CBD method is used to deposit CdS TFs along with surface vaporization and thermal evaporation techniques. The CBD is known for its poor quality usage of raw materials. The recycling capacity of the cadmium compounds are not very efficient and will leave up to 30% of cadmium as solid waste (S.V.T. Coalition, 2009). Cadmium is considered "extremely toxic" by the US Environmental Protection Agency (EPA) (Navas-Acien et al., 2004; Moskowitz and Fthenakis, 1991b; Cornelis et al., 2005) and Occupational Safety and Health Administration (OSHA) (Fthenakis et al., 1999) due to its carcinogenic nature. It can affect kidney, liver, bone, and can lead to blood damage from ingestion and lung cancer from inhalation. This can happen during the manufacturing process when the workers are exposed to cadmium compounds. The worst effects of the cadmium have controlled through the political issues. This political stand against the cadmium was implemented due to their worst health issues. The European Economic Community (EEC) has prohibited the sale of most products containing cadmium. However, the toxicology studies of cadmium telluride (CdTe) are very limited (Moskowitz and Fthenakis, 1990). Fig. 3.6 explains the key hazardous materials and their effect during manufacturing and module recycling.

The CdTe panel making in the production line creates potential dust and fumes, along with other hazardous materials used in the module making. Apart from the production line, the PV modules can release toxic elements during a fire in building integrated photovoltaics (BIPV), especially in residential and commercial buildings. However, the risk during the fire events is minimal for CdTe modules, as it holds higher temperature evaporation than typical household fires (Moskowitz and Fthenakis, 1990).

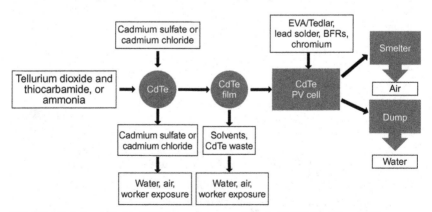

FIG. 3.6   Cadmium telluride (CdTe) generic process diagram (S.V.T. Coalition, 2009).

3.4.2 Copper-Indium Gallium Selenide (CIGS) Solar Cells

The Cu(In,Ga)Se$_2$ (CIGS) solar cells have demonstrated higher conversion efficiencies over 20% both in small devices and module level (Repins et al., 2008; Jackson et al., 2011; Green et al., 2015). CIGS is known for its long-term durability and cost-effective production methods which makes it competitive to current silicon-based PV technologies. There are different methods to fabricate CIGS solar cells such as "three-stage" co-evaporation process (Hibberd et al., 2010; Wei et al., 2014a,b) and two-stage non-vacuum chemical methods (Hibberd et al., 2010; Verma et al., 1996). The controllability of various parameters in a three-stage growth process can bring complexities for CIGS solar cell manufactures, viz. high processing temperature (500–600°C), high vacuum (10^{-6}–10^{-5} mbar), and complex controllability. But, two-step processes, which are less complicated, lead to less efficient cells. In CIGS manufacturing, selenization plays an important role to make the process complete. Selenization of metallic precursors using Se/N$_2$ (Guillén and Herrero, 2002; Kaelin et al., 2005; Wei et al., 2011) or H$_2$Se (Alberts et al., 2000; Sugimoto et al., 2011a,b) environment is considered as a simple and promising two-step approach to produce good quality absorber and high efficiency CIGS solar cells. The low temperature and non-vacuum techniques are producing poor quality absorbers. The deposition of CIGS layers requires starting materials such as copper, indium, and gallium with selenium (Se) in gaseous form. The hydrogen selenide is used in many industries and it is considered a highly toxic source of selenium. It can be very dangerous at concentrations as low as 1 ppm in the air. This is an important step to achieve the high quality TF absorber materials which is often known as selenization. The selenization is happening in higher temperatures over 575°C where it can form potential selenium dioxide (SeO$_2$) (Köppel et al., 1986; Hill, 1974; Cardwell et al., 1976; Wilber, 1980). Selenium dioxide is a tissue poison like arsenic. Selenium dioxide is vented into a water solution, where it forms elemental selenium and it can be recovered with care (Fthenakis et al., 2009). Fig. 3.7 explains the CIGS and related TF materials and its hazardous materials recycling.

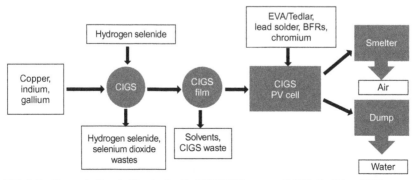

FIG. 3.7 Copper-indium (gallium) selenide (CIS/CIGS) process (S.V.T. Coalition, 2009).

3.4.3   Multijunction Solar Cells

Multijunction solar cells based on III-V materials (gallium arsenide (GaAs), aluminum indium phosphide (AlInP), aluminum gallium indium phosphide (AlGaInP), gallium indium phosphide (GaInP), and indium phosphide (InP), etc.) show high efficiency, exceeding 35%, but due to the high production cost and low availability of their constituents, these solar cells are not considered suitable for cost-effective terrestrial applications though they are still very important for space PV applications. Some of the silicon technologies (such as a-Si) have serious implications of the light induced degradation. In order to overcome limitations like light absorption, thickness of the absorber layer, and instability of the devices, the multijunction approaches have been adopted. To work within the limits of intrinsic layer thickness of $\sim$300 nm and make use of different light trapping arrangements, the concept of tandem cells using double and triple junctions has been thoroughly pursued worldwide. Multijunction solar cells are used for better utilization of solar spectrum and to improve the stability of the solar cells. GaAs is currently used in multijunction solar cells with other materials. The concentrating solar cell technology uses GaAs solar cells. The production of GaAs crystals can be through a combination of gallium and arsenic pure elements or using trimethyl gallium ($(CH_3)_3Ga$) and trimethyl arsenic ($(CH_3)_3As$) gases. Trimethyl arsenic detoxifies arsenic (Stýblo et al., 2002; Styblo et al., 2000) is rated as a poisonous material during this process. In case of accidents during the GaAs growth, it can have significant effects on lung, liver, immune, and blood systems (Webb et al., 1984, 1986; Flora and Gupta, 2013). There is little toxicological data on gallium, but it is widely used as a marker/tag in MRI tests, and believed to be safe in small doses (Table 3.4; S.V.T. Coalition, 2009).

Multijunction solar cells heavily use GaAs in their device structure. The toxic materials arsenic, phosphine, and arsine are used in manufacturing process heavily. Apart from the usual toxic materials, solvents for cleaning the TF surface and substrates used in the process also contain known carcinogens (Komeno, 1994). It is worth mentioning here that there are lots of less toxic alternatives that have also been used in this process.

3.5   NEW GENERATION SOLAR CELLS

The third-generation solar cells, such as dye sensitized solar cells (DSSC) and organic solar cells are being focused on to work with relatively less pure materials and methods to make solar cells. It makes them a cheaper alternative to silicon and TF solar cells. However, third-generation solar cells find major applications in interior and BIPV sectors, due to its stability issues. Moreover, third-generation solar cells have many associated problems, such as usage of liquid electrolyte, low stable materials, chemically unstable charge transfer, etc. Attempts have been made for many years to find an alternative to the liquid

TABLE 3.4 Hazardous Materials and Associated Health Issues in Multijunction Solar Cell Production

Hazardous Material	Environment Used	Health Issues
Arsenic	Production of GaAs crystals	Highly toxic and carcinogenic/ occupational health hazards
Phosphine and arsine	Doping with GaAs crystals	Highly toxic gases. Mostly occupational health issues
Trichloroethylene	Used as a solvent with other known solvents	Known carcinogen
Hydrochloric acid, methane, triethyl gallium, and trimethyl gallium	Used during manufacturing process	Occupational health issues

electrolytes, and thus to obtain an improved DSSC solar cell which will have ease of fabrication, less complication in the sealing, and encapsulation of the device, the possibility for monolithic interconnection of the cells within the module, and therefore also increased performance and lower cost (Goswami and Kreith, 2007). DSSC and organic solar cells are also classified as nanostructured solar cells which use lots of nanomaterials in manufacturing. These materials can generate occupational health hazards when handled. The manufacturing processes for these types of solar cells are rated as greener compared to silicon and TF technologies. Nevertheless, the system process is using acids and solvents in huge quantities to make third-generation solar cells as well.

3.5.1 Perovskite Solar Cells

Perovskite solar cells have attracted a large amount of attention due to their sudden rise in the conversion efficiency—over 20% within a short duration of time. Inorganic-organic based hybrid perovskite solar cells most commonly comprised of $CH_3NH_3PbI_3$ materials with an appropriate band gap (1.55 eV), high absorption coefficient, long hole-electron diffusion length (~100 nm), and excellent carrier transport. Moreover, perovskite solar cells are most commonly prepared through different techniques like solution process (spin coating) and thermal evaporation methods. The flexibility of making highly efficient solar cells with solid-state DSSCs will have less complication in the manufacturing process, easy possibility for producing monolithically interconnected modules, easier sealing and encapsulation of the modules which are similar to other TF solar cells. Although the efficiency values reached for small-size individual solar cells can be considered to be satisfactory, and should basically provide a great commercialization potential for this technology in the future, especially thanks to the fact that we have a true solid-state device, the stability and longevity of the device is yet to be ascertained. Critical issues are the stability of the organic spiro-OMeTAD layer and that of the perovskite with the Pb-free compounds. The lead (Pb) is one of the main issues with this type of solar cells.

Lead (Pb) is a natural metal found in many compounds in small quantities (50–400 ppm). In recent years, lead usage has been restricted due to its bad health effects, such as cardiovascular and developmental diseases to neurological and reproductive damage (Hailegnaw et al., 2015). The main source of lead comes from human activities, including burning of fossil fuels, mining and manufacturing of consumer goods such as (secondary) batteries, ammunition, soldering, piping (cf. plumbing), and electronic devices. The maximum accepted levels of lead in drinking water and air were set to 15 and 0.15 µg/L, respectively, by the US EPA (Paulose, 2014; USA Environmental Protection Agency, n.d.). The perovskite solar cells use methyl ammonium lead iodide (MAPbI$_3$), in its manufacturing process. This can be easily dissociated in the presence of highly polar solvent such as water. Normally solar panels are encapsulated very well to avoid the environment to interact. However, damage of the solar panels is a potential threat to this technology (Hailegnaw et al., 2015; Vincent et al., 1987; Zweibel et al., 1998; Frost et al., 2014). Hailegnaw et al. (2015) have reported the rain effect on perovskite solar cells and their lead emission problems. The alternative to lead is being researched to be replaced in perovskite solar cells.

3.6   PV RECYCLING HAZARDOUS MATERIALS

PV products contain not only materials for solar cells, but also have electronic waste (e-Waste). The recycling of the PV products has the challenge to recycle materials in solar cell parts and other electronics separately. The solar panels will leave the toxic materials as e-waste in landfills where they can leach into ground water and air. Table 3.5 shows the various recycle wastes from PV panels and their health issues, along with recycling options.

3.7   CONCLUSIONS

The PV-related toxicities and environmental impacts can be classified into three stages such as (i) materials level, (ii) system level, and (iii) recycling level.

1. *Materials:* The most efficient technologies like silicon, CdTe are based on processing of high-level toxic materials which have the worst effect when they are exposed to living beings or during processing. Further research into

TABLE 3.5 End Life Hazardous Materials From Different Solar Cell Systems

Type of PV	End Life Hazardous Waste	Recycling Options
Crystalline silicon (c-Si)	Lead, brominated flame-retardants, and hexavalent chromium	Melting Si wafers to make new silicon ingots
Amorphous silicon (a-Si)	Amorphous silicon panel are not containing EPA classified toxic materials	Recycled as a household waste streams due to its vast usage in electronics other than PV
Cadmium telluride (CdTe)	Cadmium and cadmium hydroxide	1. Recycled in a similar passion as cathode ray tubes (CRTs), fluorescent lights, and Ni-Cad and lead-acid batteries 2. Use of strong acids (such as sulfuric acid) to strip off metals
Copper-indium selenide (CIS) and copper-indium gallium selenide (CIGS)	Selenium, hydrogen selenide, gallium, cadmium	Indium is recycled as it is rare metal and has potential use in television screens
Gallium arsenide (GaAs) and multijunction panels	Gallium and arsenic in small doses	No particular recycling methods are in place
Nanostructured materials	Lead and potential waste from nanomaterials	No particular recycling methods are in place

recycling, utilize different materials, and environmental friendly manufacturing routes using less natural resources have to be done.

2. *System:* The engineering aspects of the PV technologies have to be refined to use minimum resources.

3. *Recycling:* This is a very important part of the PV materials to reuse the most of the resources. Material's aspect related recycling methods have to be implemented.

As a PV power plant, there are few different issues reported or observed:

1. *Reflection:* The insects view the reflection from the panels as a water body and more insects are attracted toward the PV panels. The concentrated solar cells at power plants are having different issues of higher temperature near the focal points which could kill the birds and insects. This can create more

environmental imbalance in those areas. Proper reflection mitigation plans have to be put forward to tackle this issue.

2. *Visual impacts:* The visual impact of solar power plants are minimal compared to the BIPV. However, there are research initiatives to make solar panels with pleasing colors according to the standards.

3. *Land and biological impacts:* The PV power plants are often carried over with the biological and land-related issues. However, the improvement of efficiency of the solar panels could resolve this issue.

The environmental concerns are a fear of exposure, health, and environments worsening. The more efficiency semiconductor materials are one way or the other related to toxic processing and the level of toxicity used in processing solar energy materials is staggering. Proper toxic material handling and recycling would be a way forward to make clean energy technologies into less GHG emissive and more environmental friendly. The policies in manufacturing and recycling should be properly put forward to make the technologies less toxic. This can be done through global industrial policies implications with local authorities are the way forward to make renewable energy technologies more viable.

Apart from silicon related toxicities, the perception of toxicity of constituent element Cd, which is used in the form of a stable compound in TF modules, often raised issues, such as the risks or hazards in handling CdTe technology are associated with the materials used during the processing and fabrication of CdTe/CdS and CdS/CIGS solar cells, along with the risks associated during the cradle-to-grave operating lifetime of these modules. The environmental and health hazard (E&H) issues of CdTe solar modules have been extensively investigated by several independent agencies, including the national laboratories in Europe and United States (refer Chapter 4) While CdTe technology has no chance to do away with Cd, there is some maneuverability in CIGS technology in the elimination of very thin (typically $\sim$50 nm) CdS buffer layer and so the quest for an alternative buffer layer is being successfully pursued. Initial success has already been achieved as CIGS solar cells of 16–18.8% and modules of 13.4% have been developed with alternative "Cd-free" buffers. Recently, co-evaporation of CIGS in an inline single-stage process is used to fabricate solar cell devices with up to 18.6% conversion efficiency using a CdS buffer layer and 18.2% using a $Zn_{1-x}Sn_xO_y$ (Cd-free) buffer layer. Cadmium is a by-product of zinc, lead, and copper mining. It constitutes only 0.25% of its main feedstock ZnS (sphalerite). Cadmium is released into the environment from phosphate fertilizers, burning fuels, mining, and metal processing operations, cement production and disposing of metal products. Releases from disposed Cd products, including Ni-Cd batteries are minimum contributors to human exposure because Cd is encapsulated in the sealed structures. Most human cadmium exposure comes from ingestion and most of that stems from the uptake of cadmium by plants, through fertilizers, sewage sludge, manure, and atmospheric deposition. Although long-term exposure to elemental cadmium, a carcinogen, has detrimental effect on kidneys and bones, limited data exists in toxicology. However, CdTe compound is more stable and less soluble than Cd element and therefore likely to be much less toxic.

Considering the electrolytic refinery production of CdTe powders (from Cd wastes from Zn, iron and steel industries) there would be an emission of 0.001% Cd gaseous emission. This would correspond to 0.01 g/GWh, which is significantly less as compared to the perceptions and hype created by some who estimate it at 0.5 g/GWh based on other crude processes or unsubstantiated data. The only potential hazard that could come to any one's mind would be the building fire. It has also been estimated quantitatively that the maximum temperature of a basement on fire is ∼900°C, which is still less than the melting point of CdTe (1041°C). Besides, the vapor pressure at 800°C for CdTe is ∼2.5 Torr (0.003 atm), so this minimizes the risks further and once sealed between glass plates, any Cd vapor emission is unlikely. The main conclusion of these studies was that the environmental risks associated with CdTe based technology are minimal. Every energy source or product may present some direct or indirect environmental health and safety hazards and those of CdTe should by no means be considered a problem, the following conclusions were drawn:

- Cd is produced as a by-product of Zn and can either be put to beneficial uses or discharged to the environment posing another risk.
- CdTe in PV is much safer than other current Cd uses.
- CdTe PV uses Cd 2500 times more efficiently than Ni-Cd batteries.
- Occupational health risks are well managed.
- Absolutely no emission during PV operation.
- A risk from fire emission is minimal.
- CdTe technology and modules are safe and don't pose significant risks.

The policies related CdTe issue and mitigations are discussed in Chapter 4.

ABBREVIATIONS

a-Si	amorphous silicon
Al	aluminum
AlGaInP	aluminum gallium indium phosphide
AlInP	aluminum indium phosphide
B	boron
BIPV	Building Integrated Photovoltaics
c-Si	crystalline silicon
C	carbon
Ca	calcium
CBD	chemical bath deposition
Cd	cadmium

CdCl$_2$	cadmium chloride
CdS	cadmium sulfide
CdSO$_4$	cadmium sulfate
CdTe	cadmium telluride
(CH$_3$)$_3$Ga	trimethyl gallium
(CH$_3$)$_3$As	trimethyl arsenic
CH$_3$NH$_3$PbI$_3$	methylammonium lead iodide
CIGS	copper-indium gallium diselenide
CNS	central nervous system
CO	carbon monoxide
CRT	cathode ray tube
CS(NH$_2$)$_2$	thiourea
CSS	close-spaced sublimation
CVD	chemical vapor deposition
CZTS	copper-zinc tin diselenide
DSSC	dye sensitized solar cells
ED	electro-deposition
EEC	European Economic Community
EPA	Environmental Protection Agency
Fe	iron
GaAs	gallium arsenide
GaInAs	gallium indium arsenide
GaInP	gallium indium phosphide
GaSb	gallium antimonite
GHG	greenhouse gas
GI	gastrointestinal
H$_2$	hydrogen
HCl	hydrochloric acid
InGaAs	indium gallium arsenide
InP	indium phosphide
IPCC	Intergovernmental Panel of Climate Change
LCA	Life Cycle Assessment
MG-Si	metallurgical grade silicon
Mg-Si	magnesium silicide
Mg	magnesium
Mn	manganese
MPa	megapascal
NH$_3$	ammonia
N$_2$	nitrogen
NREL	National Renewable Energy Laboratory
OSHA	Occupational Safety and Health Administration
P	phosphorus
Pb	lead
PV	photovoltaic

Se	selenium
SiCl$_4$	silicon tetrachloride
SeO$_2$	selenium dioxide
SF$_6$	sulfur hexafluoride
SF$_2$	sulfur difluoride
SiF$_4$	silicon tetrafluoride
SiHCl$_3$	trichlorosilane
SiH$_2$Cl$_2$	dichlorosilane
SiH$_4$	silane
SiHBr$_3$	silicon tetrabromide
SiO$_2$	silica
Si	silicon
SO$_2$	sulfur dioxide
Spiro-OMeTAD	2, 2',7,7'-tetrakis-(N,N-di-4-methoxyphenylamino)-9,9'-spirobifluorene
TeO$_2$	tellurium dioxide
Te	tellurium
TF	thin film
Ti	titanium
Wp	watt-peak

REFERENCES

Alberts, V., Bekker, J., Witcomb, M., Schön, J., Bucher, E., 2000. Thin Solid Films 361, 432–436.

Alsema, E., Baumann, A., Hill, R., Patterson, M., 1996. Report. Department of Science, Technology, and Society, Utrecht University, The Netherlands.

Avrutin, V., Izyumskaya, N., Morkoç, H., 2011. Superlattice. Microst. 49, 337–364.

Bosio, A., Romeo, N., Mazzamuto, S., Canevari, V., 2006. Prog. Cryst. Growth Charact. Mater. 52, 247–279.

Cardwell, R., Foreman, D., Payne, T., Wilbur, D., 1976. Arch. Environ. Contam. Toxicol. 4, 129–144.

Cha, A.E., 2008. Washington Post 9.

Cornelis, R., Caruso, J.A., Crews, H., Heumann, K.G., 2005. Handbook of Elemental Speciation, Handbook of Elemental Speciation II: Species in the Environment, Food, Medicine and Occupational Health. John Wiley & Sons, Chichester.

Fang, Z., Wang, X.C., Wu, H.C., Zhao, C.Z., 2011. Int. J. Photoenergy 2011, 1–8.

Flora, S., Gupta, S.D., 2013. Def. Sci. J. 44, 5–10.

Frost, J.M., Butler, K.T., Brivio, F., Hendon, C.H., Van Schilfgaarde, M., Walsh, A., 2014. Nano Lett. 14, 2584–2590.

Fthenakis, V.M., 1998. Prog. Photovolt. Res. Appl. 6, 91–98.

Fthenakis, V.M., 2003. Overview of potential hazards. In: Markvart, T., Castaner, L. (Eds.), Practical Handbook of Photovoltaics: Fundamentals and Applications. Elsevier, New York.

Fthenakis, V.M., 2004. Renew. Sust. Energ. Rev. 8, 303–334.

Fthenakis, V., http://www.bnl.gov/pv/abs/abs_149.asp (accessed 18.12.15).

Fthenakis, V., Alsema, E., 2006. Prog. Photovolt. Res. Appl. 14, 275–280.

Fthenakis, V., Kim, H.C., 2011. Sol. Energy 85, 1609–1628.

Fthenakis, V.M., Moskovitz, P.D., 1990. Solid State Technol. 33, 81–85.

Fthenakis, V., Morris, S., Moskowitz, P., Morgan, D., 1999. Prog. Photovolt. Res. Appl. 7, 489–497.

Fthenakis, V.M., Kim, H.C., Alsema, E., 2008. Environ. Sci. Technol. 42, 2168–2174.

Fthenakis, V., Wang, W., Kim, H.C., 2009. Renew. Sust. Energ. Rev. 13, 493–517.

Goetzberger, A., Hoffmann, V.U., 2005. Photovoltaic Solar Energy Generation. Springer Science & Business Media, Berlin.

Goswami, D.Y., Kreith, F., 2007. Handbook of Energy Efficiency and Renewable Energy. CRC Press, Boca Raton, FL.

Green, M.A., Emery, K., Hishikawa, Y., Warta, W., Dunlop, E.D., 2015. Prog. Photovolt. Res. Appl. 23, 1–9.

Gribov, B., Zinov'ev, K., 2003. Inorg. Mater. 39, 653–662.

Guillén, C., Herrero, J., 2002. Sol. Energy Mater. Sol. Cells 73, 141–149.

Hailegnaw, B., Kirmayer, S., Edri, E., Hodes, G., Cahen, D., 2015. J. Phys. Chem. Lett. 6, 1543–1547.

Hibberd, C.J., Chassaing, E., Liu, W., Mitzi, D.B., Lincot, D., Tiwari, A.N., 2010. Prog. Photovolt. Res. Appl. 18, 434–452.

Hill, C., 1974. J. Nutr. 104, 593–598.

http://www.nrel.gov/analysis/sustain_lcah.html (accessed 29.12.15).

Jackson, P., Hariskos, D., Lotter, E., Paetel, S., Wuerz, R., Menner, R., Wischmann, W., Powalla, M., 2011. Prog. Photovolt. Res. Appl. 19, 894–897.

Kaelin, M., Rudmann, D., Kurdesau, F., Zogg, H., Meyer, T., Tiwari, A.N., 2005. Thin Solid Films 480, 486–490.

Komeno, J., 1994. J. Cryst. Growth 145, 468–472.

Köppel, C., Baudisch, H., Beyer, K., Klöppel, I., Schneider, P.V., 1986. Clin. Toxicol. 24, 21–35.

Mason, J., Fthenakis, V., Hansen, T., Kim, H., 2006. Prog. Photovolt. Res. Appl. 14, 179–190.

McEvoy, A., Markvart, T., Castañer, L., Markvart, T., Castaner, L., 2003. Practical Handbook of Photovoltaics: Fundamentals and Applications. Elsevier, Amsterdam.

Moskowitz, P., 1995. Solar Cells and Their Applications. Wiley, New York, pp. 391–416.

Moskowitz, P.D., Fthenakis, V.M., 1990. Solar Cells 29, 63–71.

Moskowitz, P., Fthenakis, V., 1991a. Solar Cells 31, 513–525.

Moskowitz, P., Fthenakis, V., 1991b. Solar Cells 30, 89–99.

Navas-Acien, A., Selvin, E., Sharrett, A.R., Calderon-Aranda, E., Silbergeld, E., Guallar, E., 2004. Circulation 109, 3196–3201.

Paulose, A., 2014. Wm. & Mary Envtl. L. & Pol'y Rev. 39, 507.

Raugei, M., Bargigli, S., Ulgiati, S., 2007. Energy 32, 1310–1318.

Repins, I., Contreras, M.A., Egaas, B., DeHart, C., Scharf, J., Perkins, C.L., To, B., Noufi, R., 2008. Prog. Photovolt. Res. Appl. 16, 235–239.

Schei, A., Tuset, J.K., Tveit, H., 1998. Production of High Silicon Alloys. Tapir, Trondheim.

Stoppato, A., 2008. Energy 33, 224–232.

Styblo, M., Del Razo, L.M., Vega, L., Germolec, D.R., LeCluyse, E.L., Hamilton, G.A., Reed, W., Wang, C., Cullen, W.R., Thomas, D.J., 2000. Arch. Toxicol. 74, 289–299.

Stýblo, M., Drobná, Z., Jaspers, I., Lin, S., Thomas, D.J., 2002. Environ. Health Perspect. 110, 767.

Sugimoto, H., Kawaguchi, Y., Yasaki, Y., Aramoto, T., Tanaka, Y., Hakuma, H., Kuriyagawa, S., Kushiya, K., 2011a. In: 26th European Photovoltaic Solar Energy Conference and Exhibition, Hamburg.

Sugimoto, H., Yagioka, T., Nagahashi, M., Yasaki, Y., Kawaguchi, Y., Morimoto, T., Chiba, Y., Aramoto, T., Tanaka, Y., Hakuma, H., 2011b. In: 2011 37th IEEE Photovoltaic Specialists Conference (PVSC).

S.V.T. Coalition, 2009. Towards a Just and Sustainable Solar Energy Industry—A Silicon Valley Toxics Coalition White Paper, January 14, 2009. http://svtc.org/wp-content/uploads/Silicon_Valley_Toxics_Coalition_-_Toward_a_Just_and_Sust.pdf.

Tsuo, Y.S., Gee, J., Menna, P., Strebkov, D., Pinov, A., Zadde, V., 1998. Environmentally Benign Silicon Solar Cell Manufacturing. National Renewable Energy Laboratory, Golden, CO.

L. L. a. R. USA Environmental Protection Agency, http://www2.epa.gov/lead/lead-laws-and-regulations (retrieved 28.11.15).

Verma, S., Orbey, N., Birkmire, R.W., Russell, T., 1996. Prog. Photovolt. Res. Appl. 4, 341–353.

Vincent, B.R., Robertson, K.N., Cameron, T.S., Knop, O., 1987. Can. J. Chem. 65, 1042–1046.

Webb, D., Sipes, I., Carter, D., 1984. Toxicol. Appl. Pharmacol. 76, 96–104.

Webb, D., Wilson, S., Carter, D., 1986. Toxicol. Appl. Pharmacol. 82, 405–416.

Wei, Z., Shimell, T., Upadhyaya, H.M., 2011. In: 26th European Photovoltaic Solar Energy Conference and Exhibition, Hamburg, Germany, pp. 2840–2844.

Wei, Z., Senthilarasu, S., Yakushev, M.V., Martin, R.W., Upadhyaya, H.M., 2014a. RSC Adv. 4, 5141–5147.

Wei, Z., Bobbili, P.R., Senthilarasu, S., Shimell, T., Upadhyaya, H.M., 2014b. Surf. Coat. Technol. 241, 159–167.

Wilber, C.G., 1980. Clin. Toxicol. 17, 171–230.

Zweibel, K., Moskowitz, P., Fthenakis, V., 1998. Thin-Film Cadmium Telluride Photovoltaics: ES and H Issues, Solutions, and Perspectives. National Renewable Energy Lab, Golden, CO.

Chapter 4

The Sustainability of Solar PV Governance: A Comparative International Perspective

4.1 INTRODUCTION

The past few decades have witnessed fundamental shifts in the way social processes are governed. Here, scholars employ the notion of "governance" to denote how governments steer society towards achieving policy objectives through utilizing markets, networks of actors, and communities (eg, Pierre and Peters, 2000; Kjaer, 2005). In terms of the environment, states have traditionally employed more centralized regulatory approaches to govern, or "command and control," but in recent decades there has been a gradual movement towards including nonstate actors such as businesses and communities in deciding and implementing public policy (Benson and Jordan, 2016). This form of governing has therefore necessitated the increasing use of nonregulatory measures such as market-based instruments (MBIs), such as taxes and subsidies, voluntary agreements, and informational approaches (Jordan et al., 2012a, 2013). While MBIs seek to incentivize specific actions through economic means, voluntary agreements between governments and nonstate sectors (primarily business) bind societal actors into adopting formalized behaviors without resorting to regulation (Benson and Jordan, 2016). Informational instruments such as government guidance, meanwhile, use persuasion to influence behavior (Benson and Jordan, 2016). Such instrumental steering is evident in many environmental sectors and levels of governance (eg, Wurzel et al., 2013).

In examining the sustainability of solar photovoltaics (PV) technologies worldwide, this chapter therefore adopts a policy instruments analysis of comparative governance in several leading political contexts: the United States of America, the United Kingdom, Germany, India, and China. These states were chosen because they are leading manufacturers and/or installers of solar PV technologies, but also important in terms of their position in the global economy. Each state has taken a differing policy approach to both promoting the economic development of the solar PV technology sector and governing its environmental and social impacts, with a multiplicity of policy instruments

Solar Photovoltaic Technology Production. http://dx.doi.org/10.1016/B978-0-12-802953-4.00004-4

47

visible to support governance. In addition, each political context is comprised of multiple levels of governance, including the United Kingdom and Germany which are subject to higher level policy making in the shape of the European Union (EU). Comparing approaches between these systems can provide a basis for policy learning and potential transfer: what public policy analysts refer to as "lesson-drawing" (Benson and Jordan, 2011). To systematically analyze and compare these approaches, the chapter divides policy responses into those aimed at encouraging the growth of the solar PV sector domestically, dealing with the production-related impacts, and lastly installation and end-of-life effects identified in the previous chapter. Of interest with the latter, is whether recycling or reprocessing of waste products are encouraged or enabled. The chapter then identifies evident gaps in governance to provide some principles guiding how solar PV governance could and indeed should, normatively, evolve in the future to ensure sustainable development—features discussed in more detail in the next chapter.

4.2   SOLAR PV GOVERNANCE—UNITED STATES

Analysis of solar PV governance in the United States is necessarily complicated by the multilevel nature of its political system. As the federal state par excellence, the United States is characterized by a division of powers between the central (federal) government, state, and local levels (Watts, 2008). Each level of government is constitutionally allocated specific powers, making the governance of complex, interjurisdictional issues such as environmental problems necessarily "cooperative" (Fischman, 2005). While most environmental issues were largely enumerated to state or even local levels of government up until the 1960s, the increasing trans-boundary and chronic nature of problems led to a change in governing (see Andrews, 2006; Connelly et al., 2012). The unwillingness or inability of state governments to counter serious pollution problems led to federal government intervention. A National Environmental Policy Act (NEPA) was adopted by Congress in 1969 to provide the basis for federal environmental policy. Several landmark laws then followed, including the Clean Water Act (1972), Clean Air Act amendments (1970), the Endangered Species Act (1973), the Resource Conservation and Recovery Act (RCRA) (1976) and the Toxic Substances Control Act (1976). These laws provide the framework for controlling environmental impacts across states, although they are responsible for implementation under federal government oversight provided by the Environmental Protection Agency (USEPA). They also frame how some impacts from PV technologies are controlled. The federal government has also actively supported US energy production through policy, giving it a pivotal role in supporting solar PV technologies.

4.2.1   Sector Development Policies—Federal and State Levels

As a consequence of this multilevel system, federal, state, and local policy is influential in the expansion of the sector. By 2015, the total US capacity for installed solar power was 17.55 GW, with nearly 5 GW added in 2013 (Deutsche Bank, 2015, p. 62). Increasingly, this installed capacity is in the form of distributed generation, with many states, cities, and communities "experimenting with policies to encourage distributed solar to offset peak electricity demand" (SEIA, 2015a). A variety of multilevel policy instruments have been employed by governments and nonstate actors to encourage this polycentric growth in domestic capacity (Table 4.1).

At the federal level, government policy has utilized regulation and MBI incentives, mainly through the tax system. The US government was an early exponent of regulating to promote renewables, via the Public Utility Regulatory Policy Act of 1978 (PURPA). The Act compelled public utilities to source more

TABLE 4.1 Main Federal Level Solar PV Policy Instruments for Sector Development in the USA

Instrument	Instrument Type	Year of Adoption	Main Aims
Investment Tax Credit (ITC)	MBI	2005	To provide tax exemption for PV installation costs via the federal tax code
Modified Accelerated Cost Recovery System (MACRS)	MBI	1986	To provide tax exemption for capital costs over a five-year period from installation
1603 Treasury Program	MBI	2009	To provide federal grants for solar projects
Loan Guarantee Program (DOE)	MBI	2005	To provide loan guarantees for solar installation and manufacturing
Clean Energy Contracting	Regulation	2005	To compel federal agencies to purchase renewable energy
PURPA	Regulation	1978	To compel public utilities to source energy supplies from small producers

cost-effective supplies from smaller independent power production facilities (under 80 MW), also promoting alternative (nonfossil) production. Market steering has been primarily achieved through the federal solar Investment Tax Credit (ITC). First introduced under the Energy Policy Act 2005,[1] the ITC provides partial taxation exemption under the federal tax code for residential and commercial properties to install solar powered systems (US Department of Energy, 2015a). These can include systems for electricity generation, heating and cooling, but not passive solar or solar pool-heating. Individuals or companies can claim up to 30% of their tax liability through offsetting it against the costs of expenditures. After 2016, the tax credit will expire but, given its popularity, Congress may vote to extend it. Other income tax deductions can be made under federal rules for the recovery of capital costs for business investment in solar projects via the Modified Accelerated Cost Recovery System (MACRS) (US Department of Energy, 2015b). Businesses are allowed to deduct capital costs from their tax liability, so-called depreciation deductions, annually over a five-year period from installation, thereby providing more rapid returns on investment. Meanwhile, other innovative fiscal instruments such as the 1603 Treasury Program are also stimulating distributed solar PV installation (US Treasury, 2015). Adopted under the American Recovery and Reinvestment Act 2009 (see Russel and Benson, 2014), the program is aimed at leveraging finance for installing solar projects via a federal Treasury grant award. By waiving their eligibility for tax credits under the ITC, installers of commercial solar technology can claim grant funding directly from the government. Federal policy, introduced under Section 1703 of the Energy Policy Act 2005, also offers support for investors via the Department of Energy Loan Guarantee Program, which underwrites commercial loans for renewable installation, but also manufacturing as well. Issued by the Loan Programs Office, the grants have provided over $4.6 billion in loan guarantees for constructing utility-scale (100 MW plus) projects, which according to the Department have "helped transform U.S. energy production and paved the way for the fastest growing sector of the solar industry" (US Department of Energy, 2015c, p. 6).

The federal government has employed limited regulatory approaches in addition to MBIs as part of solar industry development policy. Under the federal Energy Policy Act 2005, Congress established a legal requirement for all federal agencies to derive 5% of their electricity consumption from renewable resources between 2010 and 2012, rising to 7.5% thereafter (SEIA, 2015a). While theoretically this legal requirement for Clean Energy Contracting should provide strong support for solar generation, given the significant energy requirements of federal government bodies,[2] it is limited by the time constraints

---

1. Amended by the Energy and Extension Act 2008 and the American Recovery and Reinvestment Act 2009.
2. According to the SEIA (2015a), the US government is the nation's largest utility purchaser, with annual electricity costs of $5.8 billion.

imposed upon Power Purchase Agreements (PPAs) that agencies can agree with suppliers (SEIA, 2015a). Other regulations prevent federal agencies, apart from the Department of Defense, from signing PPAs for periods longer than 10 years, which presents disincentives for some renewable projects such as solar where payback times can be long.

In the United States, multiple policy instruments exist at the US state and local level. Both regulatory and MBI approaches are widespread, but variable in their application. Over half of states in 2013 had set Renewable Electricity Standards (RESs) or Renewable Portfolio Standards (RPS) (SEIA, 2015b). These regulatory instruments compel utility companies to market a specified percentage of their generation and/or supply from renewables, which can include so-called carve-outs (mandatory targets) for solar power. The highest RESs are set by Hawaii (40% by 2030) and California, which adopted a 33% RPS in 2011 (DSIRE, 2015). The Standard requires amendments to the California Public Utilities Commission's Energy Resources Program to ensure that one-third of all retailed electricity is produced from renewables by 2020. Other states have established portfolio goals for renewables that are nonbinding. Another key instrument for promoting renewable generation is Solar Renewable Energy Certificates (SRECs). While they vary according to state context, in essence SRECs involve creating a certificate for each 1 MWh of solar generated electricity which can then be sold on a market to utility companies (Deutsche Bank, 2015, p. 73). Utility suppliers must obtain a certain percentage of such certificates, according to the figure set by the states' RES. This type of MBI has been highly successful in supporting small-scale distributed solar generation, as it provides a market incentive for producers to increase generation. Several states, including New Jersey, Maryland, and Delaware, all have functioning markets for SRECs, while California operates a tradable renewable energy credit (TREC) market that differs significantly from this model (Deutsche Bank, 2015). Other policy instruments that operate at state level include solar tax exemptions on property and sales tax (see DSIRE, 2015). For example, the SEIA (2015c) describes how Arizona exempts retail purchases and installation of solar energy systems from state sales tax. Other states allow local authorities to provide exemptions from property taxes if solar energy systems are installed (SEIA, 2015c). Reverse auction mechanisms are also employed in California to allow distributed producers of solar electricity to sell their output to utilities. Here, producers state a specific price which they are willing to accept for their generation that can then be purchased by utilities on an open market. California also provides financial incentives for distributed self-generation of solar power under several schemes: the California Solar Initiative, Net Energy Metering, and the Self-Generation Incentive Program (see CPUC, 2015). Net Energy Metering, for example, allocates generators credits for each unit of solar power generated which then can be used to reduce energy bills from public utilities.

4.2.2 Solar PV Production Impact Policies—Federal and State Levels

The United States produces relatively limited amounts of solar PV systems: only 4% of the global total (Yue et al., 2014, p. 670). However, it is home to several leading solar module and cell manufacturing companies meaning that there are some potential impacts that require governing.

Production of solar panels in the United States is subject to specific federal regulatory standards. Discharges to air and water from industrial plants are tightly regulated under the federal Clean Air and Clean Water Acts, while wastes are subject to the RCRA. For example, silicon tetrachloride, a by-product of polysilicon used in solar PV production, is now recycled by US manufacturers, because its release to the environment is highly regulated (Nath, 2010). In 1986, the Silicon Valley Toxics Coalition (SVTC), based in California, campaigned with other environmentalists nationwide to persuade Congress to pass the national Emergency Planning and Community Right-to-Know Act, which established toxic chemical reporting requirements for states, local governments, and industry. The requirements of the act include the Toxics Release Inventory (TRI)—an inventory of routine toxic chemicals emissions required of facilities with 10 or more employees that use 10,000 pounds or more of Environmental Protection Agency (EPA)-listed hazardous chemicals. In addition, companies are required to prepare Materials Safety Data Sheets (MSDSs) that list each chemical's common name and health impacts. These are used to inform workers and emergency response teams of potential risks.

Solar PV manufacturing is subject to the requirements above if the materials used are EPA-listed. However, there is concern that new nanomaterials used in solar panels will not be adequately covered by these regulations because existing reporting requirements and regulations are based on the volume of materials in use. Although nanomaterials often exhibit very different properties than larger-sized particles of the materials, they are considered the same for regulatory purposes. For example, some nanomaterials have the potential to be extremely toxic in very small amounts, but current regulations do not reflect this danger. Therefore, current MSDSs for bulk materials may not adequately address the potential hazards of nanomaterials.

Aside from these measures, the sustainability of solar production is primarily left to industry sponsored voluntary and informational schemes, discussed further in Chapter 5.

4.2.3 Solar PV Installation and End-of-Life Policies—Federal and State Levels

Installation of solar PV systems is subject to some state legislation and policy. The environmental impacts of large development projects are covered by federal legal requirements (NEPA) to conduct an environmental assessment, but these only apply to actions undertaken by federal agencies. Individual states also require environmental assessments of large development projects. For example, California adopted the state Environmental Quality Act (CEQA) in 1970 which compels state and local agencies to account for the environmental impacts of development projects in their decision-making. An initial study (review) of the project must be undertaken, alternatives considered, and then an environmental impact report (EIR) produced. Large-scale solar projects have been subject to the CEQA, although in 2011 exemptions were provided for roof-top and parking lot projects under statute SB226.

No federal government mandated solar panel recycling schemes exist in the United States, hence recycling or disposal of PV panels is regulated under national waste management and hazardous waste laws. In this respect, the federal RCRA is the main legislative instrument for managing end-of-life solar PV panels. The RCRA sets out procedures for the disposal and recycling of waste material (EPA, 2015). A critical factor determining whether panels are subject to the regulations is the toxicity of materials they contain. A Toxicity Characteristic Leach Procedure (TCLP) must be conducted on new solar panels to assess the degree of hazard (EPA, 2012). If toxicity levels fall below certain criteria specified under the regulations, solar panel waste can be accepted at processing facilities. But if the criteria are exceeded, the waste products must be classified as hazardous and end-of-life waste processed accordingly, that is, at hazardous waste processing facilities. Decommissioned or defective solar panels are currently considered hazardous waste by regulators if they do not meet the US Environmental Protection Agency (EPA) Toxicity Characteristic Leaching Procedure (TCLP) standards (and this determination varies depending on the technology used). TCLP is intended to ensure that potentially toxic materials do not leach into the groundwater near waste disposal sites. The TCLP test is required for all new solar panels that enter the US market. If solar panels are determined to be hazardous waste, RCRA could be used to regulate their handling, recycling, reuse, storage, treatment, and disposal.

Approaches differ at the state level, where authorities can legally impose higher standards than federal regulations. California implements the Hazardous Waste Control Law (HWCL) 1972, along with other state policy on disposal of toxic materials. A dedicated agency, the California Department of Toxic Substances Control (or DTSC) is responsible for implementation. The HWCL has even stricter hazardous waste designations than the federal government, requiring that materials pass an additional toxicity test. California is the only US state with a toxics policy similar to the EU.[3] For example, California prohibits the manufacture, distribution, and processing of brominated diphenylethers (PBDEs) and also requires that substitutes for brominated flame retardants (BFRs) not be persistent, bioaccumulative, or toxic.

---

3. The E.U. relies on the similar German "DEV S4" (Deutsches Einheitsverfahren) test.

4.3 EUROPEAN UNION, THE UNITED KINGDOM AND GERMANY

It is problematic to examine solar PV governance in European states without reference to its multilevel context, more specifically EU policy. Since the early 1970s, the European Economic Community (EEC) and its successor, the EU have adopted a broad *acquis communautaire* or corpus of environmental and energy policy (Benson and Jordan, 2013). As the "regulatory state" (Majone, 1994), the EU relies on regulations to achieve its environmental objectives. Despite the EU's increasing use of nonregulatory policy instruments (Jordan et al., 2012b), "command and control" and the precautionary principle still remain the principal approaches to governing production and end-of-life impacts of PVs. However, at the national level, in the United Kingdom and Germany, a variety of policy instruments for promoting development of the PV sectors are visible. The following section will therefore initially outline the EU level context to PV governance, before examining policy in both of these countries.

4.3.1 Sector Development Policies—EU and Member State Levels

As a global climate policy entrepreneur, the EU has introduced multiple policy instruments to meet its declared commitments for climate emissions mitigation. In 2008, Member States agreed a target of 20% emissions reductions by 2020—the so-called 20–20–20 agreement. Meeting this commitment has led to several policy responses, including a revised Renewable Energy Directive (2009/28/EC) which sets out a framework for the development of the renewable energy sector in the EU. It establishes a legal compulsion on the EU to meet 20% of its energy consumption from renewables by 2020. Meeting the target is addressed through the principle of differentiated responsibilities. Each Member State must introduce a national action plan for developing renewable energy, with targets differing according to the national context. Wealthy states, such as Sweden have a high target (49%) while poorer states, such as Malta much lower (10%). Progress is measured every 2 years, with Member States compelled to produce reports on implementation. Although the national plans do not need to specifically promote solar power, Member States are encouraged to include different renewable sources in their overall energy "mix." While overtly a regulatory response to promoting renewables, the EU also employs other policy instruments, including financing solar projects via institutions such as the European Investment Bank (EIB). Although the EU is unable to grant direct tax subsidies in national contexts, the EU Directive 2003/96/EC, does allow Member States to introduce tax exemptions for renewables, including solar, where they do not distort market competition. The European Commission also distributes advice on how national plans can best meet targets through the introduction of specific instruments, primarily MBIs, such as feed-in tariffs (FITs) for renewable producers (C(2013) 7243 final). Impacts of this policy framework have proved variable, with some countries such as Germany already hugely successful in promoting its solar industry. Others such as the United Kingdom are only beginning to expand this sector.

4.3.2 Germany

Germany is a global leader in installed solar power generation. Rapid expansion of the industry pre-2013 has now declined since regulatory changes. According to the EPIA (2014, p. 24), Germany had achieved a cumulative total of 35.7 GW of solar PV installations by 2013 out of a total of 80 GW in Europe. By the following year, it had risen to 38.5 GW, with PV-generated power providing around 6.9% of German electricity consumption (Fraunhofer Institute, 2015, p. 5). Indeed, Germany accounted for 30% of the European PV installation market in 2013 (Fraunhofer Institute, 2015, p. 25). A major factor in driving this growth has been federal policy as part of the national *Energiewende*, or energy transition that aims to reduce and eventually replace both nuclear and fossil-fuel power.

Germany has employed a combination of regulations and MBIs within federal solar policy. A key mechanism is the Renewable Energy Act (REA), which was last amended in 2014. Adopted in 2000, the EEG requires large energy companies to purchase supplies from smaller independent producers, including renewables suppliers. Such producers are paid by a fixed FIT for a period of 20 years, set by government. The additional costs of solar power are then passed on to customers in their electricity bills. Controversially, the premium has been paid by homeowners, but not industry, which receives an exemption. In the past, the level of the tariff for roof and ground mounted solar systems was generous, stimulating a significant expansion in the industry, but it was scaled back in 2014 as the costs to consumers rose. Under the changes to the Act, which were effective from 2015, the FIT has been reduced to around 8–12.5 cents/kWh from an average of 32 cents/kWh in 2013 (Fraunhofer Institute, 2015, pp. 10–11). Government policy aims at phasing out the tariff by 2020, by which point it is estimated that solar PV generation will achieve cost parity with other sectors such as wind power and fossil-fuels.

4.3.3 United Kingdom

Under the RED, the United Kingdom is obliged to achieve 15% (up from 3%) of its energy consumption from renewable sources by 2020. Although still small in comparison to Germany, the UK solar sector is growing rapidly, with 3.3 GW installed in 2013 (EPIA, 2014, p. 34). Several national policy instruments are driving this expansion (Table 4.2). Long-term national policy is driven by

TABLE 4.2 Main National Level Solar PV Policy Instruments for Sector Development in the United Kingdom

Instrument	Instrument Type	Year of Adoption	Main Aims
Renewables Obligation (RO)	Regulatory	2002	To compel UK electricity suppliers to purchase renewable energy
Contracts for difference (CfD)	MBIs	N/A	To guarantee prices for renewable electricity
Feed-in tariffs (FITs)	MBIs	N/A	To provide financial incentives for small-scale solar electricity generation
Renewable Heat Incentive (RHI)	MBI	2011	To provide financial incentives to consumers to install renewable heating technology including solar thermal panels

the Climate Change Act 2008 that commits the United Kingdom to an 80% reduction in greenhouse gas emissions by 2050 (Lorenzoni and Benson, 2014). In turn, this measure is implemented by the UK Renewable Energy Roadmap, which dictates that 30% of electricity should be generated by renewables, with solar playing an important part.

The main regulatory mechanism is the Renewables Obligation (RO), introduced in 2002. Under its requirements, UK electricity suppliers must source a set level of power supplied to customers from renewables. This "obligation" is recalibrated and increased on an annual basis. Generators of renewable electricity receive Renewable Obligation Certificates (ROCs) from the government industry regulator Ofgem according to their output. They can then sell their Certificates to electricity suppliers, thereby establishing a market trading system. Suppliers must also demonstrate to Ofgem that they have purchased enough ROCs to meet the required level. Failure to do so incurs a financial penalty or "buy-out." The ROC now only applies to solar schemes under 5 MW as the success of the original mechanism led to larger installations (above 5 MW) being withdrawn from it. They are now considered under the contracts for difference (CfD) instrument that guarantees a fixed price for electricity generated, but must compete for such support with other large renewable schemes such as wind farms. FITs are an MBI used by the government to promote low-carbon electricity generation below 5 MW installed capacity systems such as solar PVs. The scheme supports individuals or organizations who install small-scale electricity generation such as solar panels. Generators must register their installation with a licensed supplier who then "pay" a tariff according to the electricity generated, with the licensee recovering the costs through reductions in the generators electricity bills. The scheme has proved popular with homeowners as they can install solar PVs on their properties and "export" any excess electricity not required for domestic use to the local power grid. Also at the domestic level, the Renewable Heat Incentive (RHI) was introduced in 2011. It is an MBI that provides financial incentives to consumers to install renewable heating technology, including biomass fueled boilers and solar thermal panels. Tariffs on each unit of heat generated are paid to consumers on a quarterly basis, with payments calculated in pence per kilowatt-hour. Initially the scheme only made payments to nondomestic business and public sector organizations but was subsequently extended in 2014 to include private individuals. Government also makes use of informational instruments and has produced several microgeneration guides in with conjunction the Energy Saving Trust.

4.3.4   Solar PV Production Impact Policies—EU and Member State Levels

Production of PV modules in the EU is at a relatively small-scale compared to China (see below): around 10% of total global production (Yue et al., 2014). While Germany manufactures modules and cells, this sector is negligible in the United Kingdom. At the EU level, industrial production impacts are primarily regulated under the concept of integrated pollution and prevention control (IPPC). Although several countries historically sought to regulate emissions from industrial processes in a holistic manner, integrated pollution control was first adopted by the EU in 1996 (Directive 96/61/EC). Now updated as the Industrial Emissions Directive (IED) 2013, it compels the site-specific permitting of industrial facilities in EU Member States to control their emissions to different environmental media. Implementing national authorities must register facilities undertaking specified production processes and issue integrated pollution permits. Permits must cover all emissions from plants according to Emission Limit Values (ELVs) set by the EU. Operators must also mitigate emissions by employing Best Available Techniques (BAT) solutions, implemented through the use of BREFs (BAT Reference guidance documents) which again are established by the EU.

Another important regulatory instrument controlling pollution from industrial production is the Restriction of Hazardous Substances (RoHS) Directive 2003, as amended, although at this point it does not cover PVs. It restricts the use of specific hazardous materials in electronic and electrical manufacturing processes. In this respect, several substances are restricted in terms of their use, namely: lead (Pb), mercury (Hg), cadmium (Cd), hexavalent chromium (Cr^{6+}), polybrominated biphenyls, and polybrominated diphenyl ether

(PBDE). However, an exemption for PV manufacturing was made in the original legislation and has since been extended. For the moment, Article 2(4)(i) of the Directive states that its obligations do not apply to "phtotovoltaic panels intended to be used in a system that... produce[s] energy from solar light..." (Official Journal, 2011). In anticipation of the inclusion of PV systems in waste electrical and electronic equipment (WEEE) legislation, the European Photovoltaic Industry Association and German Business Association launched the PV Cycle program to develop a European-wide collection, recycling, and recovery system. This effort by industry (although only voluntary) is an excellent first step in minimizing the end-of-life impacts of PV systems, but should not preclude the specific inclusion of solar PV systems in the WEEE.

Permitting practice for IPPC varies slightly between Member States. Some countries such as Germany integrate pollution permits with land use planning permission. As competences for environmental protection are shared between the federal government and Länder (federal states), implementing regulations exist at both levels. The main federal law concerning industrial permitting is the Federal Emission Control Act (Bundesimmissionsschutzgesetz, BImSchG) and related ordinances. Under the BImSchG, which implements the IED, large industrial facilities are legally required to obtain an emission control permit. Guideline BREF documents for implementing the regulations are published by the Federal Environmental Agency. However, under the German federal system, the Länder are responsible for implementing the BImSchG in their jurisdiction. Permitting therefore differs slightly between Länder: in some a state-level agency is responsible for the licensing, in others it is the local level environmental agencies of districts or towns. However, the respective environmental agency is the competent licensing authority responsible for the emission control permit procedure under the BImSchG and associated ordinances. A formal authorization procedure is conducted for specified industrial installations, including energy production and manufacturing that require an environmental impact assessment (EIA). The approval process can be considered under a "simplified procedure" for facilities with limited environmental impacts and an EIA is not required.

The United Kingdom implements EU industrial pollution legislation through national implementing regulations, although these differ slightly according to national context, due to the UK's devolved political system. In England and Wales, the regulatory framework is determined by the EPR 2010 SI2010 No. 675, as amended. Known more commonly as the Environmental Permitting Regime or EPR for short, the regulations oblige industrial operators to acquire a permit for certain specified activities or register an exemption from permitting. These activities are identified in Schedules to the regulations which also place specific requirements on operators. Permits are issued separately by the national Environment Agency, which is responsible for inspection of industrial sites. Similar governance arrangements exist for implementation in Northern Ireland and Scotland.

4.3.5 Solar PV Installation and End-of-Life Policies—EU and Member States

The installation of solar PV systems is subject to EU, national, and local level regulation, depending on the scale of installation. EIA is required for energy projects under Directive 85/337/EEC (as amended by Directive 97/11/EC). Projects must be assessed for their potential environmental impacts as part of the planning process, with impacts such as cultural heritage, landscape, and habitats assessed. Specific categories of industrial development are subject to mandatory EIA (under Annex 1) while other developments may require assessment depending on their predicted impacts (Annex 2). Generally, large-scale energy projects are subject to an EIA.

As a directive, the EU legislation allows some flexibility in implementation. EIA therefore varies between Member States, depending on the planning system. In Germany, solar projects are considered under several federal acts, including the Federal Regional Planning Act which mandates a regional planning procedure to consider the impacts of large-scale solar parks. Under federal planning legislation, a formal permission procedure is conducted for specified installations, such as wind farms and solar projects that require an EIA.[4] The assessment consists of the production of an environmental report on potential impacts. In addition, the federal Building Code (BauGB) requires an environmental report, which includes an impact assessment, a protected species assessment, and impact mitigation.

In the United Kingdom, national planning policy provides a framework for local level implementation, although planning authorities have some decisional discretion. National Planning Policy Framework (NPPF) sets out requirements for considering renewable energy projects. It instructs local authorities to adopt a positive strategy for renewable energy generation. There is also a legal requirement for conducting EIA under the national EIA Regulations 1999. Although large solar arrays are not specifically listed in Schedule 2 of the Regulations, which determines projects that may be subject to EIA, they could be subject to its requirements depending the potential impacts of the development. If EIA is required, several criteria of impact may be considered, including the effects on habitats, soils, landscape, cultural heritage, and water resources. Planners are also compelled to consider the cumulative impacts on landscapes of multiple developments. Different regulations apply to small-scale (ie, roof-mounted) PV systems, which are considered under general planning permission procedures by local authorities on a case-by-case basis. Examples of solar projects subject to this process include the Wheal Jane Mine in Cornwall (BRE, 2013). An application was submitted by developers to install a 1.55 MW solar farm on this former tin mine consisting of 5760 solar panels. The application

---

4. These requirements, under the federal Act on the Assessment of Environmental Impacts (1990), implement the EU EIA Directive.

was initially screened for EIA under the regulations and then assessed. Significant issues identified included visual and landscape impacts, in addition to ecological effects. After mitigation measures were adopted in the application, it was granted and the site became operational in 2011.

End-of-life processing of solar PV panels across the EU is subject to several regulatory measures. The Waste Framework Directive (2008/98/EC) lays down the basic principles of waste management, which have implications for PV disposal: most notably, the Polluter Pays Principle and the waste hierarchy. Measures are also established under the Directive for the disposal of hazardous wastes. More specifically related legislation is contained in the Waste Electrical and Electronic Equipment Directive (WEEE) (2002/96/EC) that regulates the collection, treatment, and disposal of electronic products while also placing some restrictions on their design. An important feature is the requirement on manufacturers of such products is to establish mechanisms to collect end-of-life waste and recycle it. WEEE provides some flexibility in national implementing procedures meaning that collection approaches vary in different countries. Solar products were originally exempt from the directive, causing some controversy. However, after recent legal amendments in 2012, solar PV units are now included in the requirements, meaning producers will have to collect and recycle their products when they become redundant. Additional restrictions on the exportation of waste from the EU are imposed by other directives and the Basel Convention.

Waste management in Germany is tightly regulated. The Waste Management Act (KrWG) provides the main regulatory framework for waste disposal. Disposal of specific wastes is covered by specific ordinances. End-of-life electrical and electronic wastes are subject to the conditions set out in the Elektro- und Elektronikgerategesetz (ElektroG).

The national Waste Electrical and Electronic Equipment (WEEE) Regulations 2013 implement the WEEE Directive in the United Kingdom. They apply to items "placed on the market" in the United Kingdom according to a definition contained in the Regulations. Also, they must fall into 1 of 10 categories listed in a Schedule. In accordance with the revised EU directive, UK Government guidance (UK Department of Business, Innovation and Skills, 2013, p. 4) states that "[f]rom Jan. 1, 2014 photovoltaic panels will come in to the scope of the Regulations for the first time." Schedule 2(4) therefore refers to "consumer equipment and photovoltaic panels" as requiring collection and recycling. The implementing authorities in the United Kingdom are the Environment Agencies of each country. To facilitate collection of waste, all producers "placing 5 tonnes or more of EEE onto the UK market" annually must join a Producer Compliance Scheme (PCS) (UK Department of Business, Innovation and Skills, 2013, p. 12). National implementing authorities must approve all PCSs. PCS arrange for WEEE to be collected and treated/reused at an Approved Authorised Treatment Facility (AATF) or Approved Exporter (AE). EEE must be treated in line with guidance on Best Available Treatment,

Recovery and Recycling Techniques (BATRRT). Producers demonstrate compliance through registration with a PCS. Critically, there are obligations on "distributors" of WEEE, as defined under Regulation 2 of the WEEE Regulations. In addition, "Distributors of EEE who also place EE onto the UK market (including manufacturing, re-branding, or by importing on a professional basis) are also classified as a producer of EEE and will have additional responsibilities under the Regulations" (UK Department of Business, Innovation and Skills, 2013, p. 19). These obligations include establishing collection schemes for specific items. AEs of such wastes must also be approved.

4.4   INDIA

India's rapidly growing solar sector has been encouraged by policy mechanisms at the national (federal) and state levels. By 2014, national installed capacity had reached around 2.8 GW, of which circa 2 GW was installed in the preceding 2 years (Deutsche Bank, 2015, p. 105). The main federal policy is the Jawaharlal Nehru National Solar Mission (JNNSM). Launched in 2010, the policy has targeted the installation of 20 GW of solar generation capacity nationally by 2022. Achieving this target has involved providing tariff subsidies and reducing costs to achieve grid parity. India also has imposed a renewable energy obligation that includes a commitment to solar.

4.4.1   Sector Development Policies—Federal and State Levels

The JNNSM policy has three phases spanning 10 years and is implemented by a specific institutional framework. Phase 1 (2012–13) is aimed at installing 1 GW of solar projects, followed by 9 GW in Phase 2 (2013–17), and 10 GW in Phase 3 (2017–22) (Deutsche Bank, 2015). A new Ministry of New and Renewable Energy (MNRE) was established to help implement the policy with a remit for production, development, and application of solar power (Murthy, 2014). It oversees the Indian Renewable Energy Development Agency Ltd (IREDA), which provides financing for electricity generation projects and solar technology manufacturers, and the Solar Energy Corporation of India (SECI) that supports the Ministry in implementing the policy. Other important government institutions are the federal Ministry of Power, the National Thermal Power Corporation, the Central Electricity Regulatory Commission, and the Central Electricity Authority. The Ministry of Power is responsible for power production in India and hence, the overall completion of the Solar Mission, in conjunction with the MNRE.

Several policy instruments are employed in meeting national solar objectives (Table 4.3). Significant regulatory measures include the Electricity Act 2003, which mandates that State Electricity Regulatory Commissions provide grid connections for renewable generation and oblige power companies to purchase a specific amount of such generation (Sharma et al., 2012). A National

TABLE 4.3 Main National Level Solar PV Policy Instruments for Sector
Development in India

Instrument	Instrument Type	Year of Adoption	Main Aims
Renewable Purchase Obligation/ Renewable Energy Certificate	Regulatory and MBI	2003	To compel energy suppliers to purchase renewable energy and provide consistency across states
Bundling scheme	MBI	N/A	To provide a mechanism for the government to sell renewable energy, "bundled" with conventional sources, at a subsidized rate
Generation-based incentive	MBI	2009	To provide a price-based incentive for smaller PV projects
Viability gap funding	MBI	2004	To provide capital funding for project installation of solar projects not anticipated to have high financial returns
National Tariff Policy	MBI	2006	To compel states to adopt a Renewable Purchase Obligation

Tariff Policy (NTP) 2006 also provides consistency across India between electricity generation and supply to customers in different states. Under its requirements, state regulatory bodies are made responsible for determining a Renewable Purchase Obligation (RPO). As in the United States, distribution companies in individual states are required to purchase a proportion of renewable electricity, classified as solar and nonsolar. Set on a state-by-state basis, the RPO is implemented by a Renewable Energy Certificate approach. Generators are issued with a certificate for every 1 MWh of electricity fed into the grid, which is then purchased by utilities on a competitive open market, or Power Exchange, to sell on to customers. Trading is overseen by government agencies. Murthy (2014, p. 11) describes how various subsidies are also used to promote investment in solar projects: a bundling scheme in which the government purchases power from project developers, "bundles" it with conventionally produced energy and sells it at a subsidized rate; a generation-based incentive (GBI) designed for smaller (100 kW–2 MW) projects; and viability gap funding for the capital costs of solar projects, selected through open competitive bidding, predicted to have low financial returns.

As in the United States, India is a federal country, meaning that states policy is highly influential in implementation of the Solar Mission, although measures differ between contexts. Rajasthan has adopted multiple policy measures to implement the JNNSM (see Ummadisingu and Soni, 2011). The Rajasthan government adopted its Solar Energy Policy in 2011, although it is now under review. It aims to install 25 GW over the next 7 years, through state or private enterprises or public-private partnerships (Government of Rajasthan, 2014). Meeting this target received a significant boost in 2014 when a Memorandum of Understanding was reached between industry and the government to install 5 GW of solar power generation (Deutsche Bank, 2015). Tamil Nadu adopted its state policy in 2012. The state is considered highly suited to solar energy production due to its insolation levels: 5.5–6 kWh/m^2/day (Government of Tamil Nadu, 2012, p. 8). A key element of state policy is the Solar Power Obligation, a regulatory instrument that obliges consumers, including businesses, but not individuals, to purchase 6% of their electricity from solar through acquiring renewable energy certificates. Rooftop generation is being promoted through a GBI. Other measures include installation of solar systems on public buildings and promoting solar heating, along with several MBIs in the form of tax concessions for solar projects. The Government of Karnataka published its revised solar policy in 2014, aiming to install 2 GW of capacity by 2021. A variety of different fiscal instruments, primarily subsidies and market mechanisms, have been adopted under the policy (see Government of Karnataka, 2014, p. 13). Punjab is another state where solar power is being heavily promoted. The Indian government has announced plans to construct a 2 GW plant in Punjab, while the state government is strongly supporting rooftop PV installation. Meanwhile, Andhra Pradesh state has adopted a variety of policy instruments for expanding the solar sector (Government of Andhra Pradesh, 2015).

4.4.2 Solar PV Production Impact Policies—Federal and State Levels

Despite attempts by the Indian government, the solar manufacturing sector in India is still in its infancy. By 2013, Sahoo and Shrimali (2013, p. 1471) record that Indian crystalline silicon cells and modules accounted for only 5% of global capacity, with less than 2% of thin-film capacity. They argue that India has a "virtual absence at the upstream [manufacturing] steps," suggesting that this sector has failed to innovate and expand (ibid.). A national Semiconductor Policy was adopted in 2007 in order to support the nascent electronics manufacturing industry through capital subsidies (Sharma et al., 2012). A domestic content requirement (DCR) was also included in the JNNSM policy, stipulating that first phase projects using crystalline silicon technologies should employ Indian manufacture modules, and that in the second phase, both modules and cells should be manufactured domestically (Sahoo and Shrimali, 2013). But neither policy helped to stimulate industrial growth in this sector. Specific problems have arisen with high dependency on imported components for cell manufacture, lack of affordable finance, competition from China and Taiwan, and limited technical capacity (Khare et al., 2013). Predictions are that the DCR policy is unlikely to stimulate significant future growth without amendments (Sahoo and Shrimali, 2013).

As a result, there are currently few evident environmental impacts from solar PV production. National environmental legislation in this area is, however, weak. The Environmental Protection Act 1986, introduced after the Bhopal industrial accident, provides a broad framework of legal principles and obligations that frame more specific pollution legislation in the form of the Water and Air acts. Fines are imposed for noncompliance. However, pollution control is primarily implemented by state governments. State Pollution Control Boards (SPCBs) are designated as the competent authorities for issuing both consents to establish (CTE) and consent to operate (CTO) for solar projects (MNRE, 2013). Despite a constitutional commitment to environmental protection and the legal framework of the national Act, fines are negligible and rarely implemented (Panigrahi and Amirapu, 2012). The government has also legislated to reduce toxic materials employed in electronics manufacture. As in the EU, the Indian government compels a reduction in the use of hazardous substances (RoHS) for electrical and electronic equipment via its E-Waste Rules, discussed below.

4.4.3 Solar PV Installation and End-of-Life Policies—National and Provincial Levels

A CTE can only be granted after an environmental assessment is conducted, although under Indian regulations only solar PV facilities above 50 ha are listed under projects requiring mandatory EIA contained in the Environmental Impact Assessment (EIA) Notification 2006. Development projects funded by overseas donors such as the World Bank may be subject to additional assessment procedures. In addition, projects have to consider the rights of indigenous populations under other federal legislation. According to Panigrahi and Amirapu (2012) India has a sound EIA legal framework but weaknesses exist in its institutional structure, implementation, and practice. As in many EIA systems globally, public participation in the assessment process is weak, with only one public hearing conducted and generally little consideration of discussions in final decision-making (Panigrahi and Amirapu, 2012).

Waste management of electronic material, or e-waste, is also a problem area for Indian governance. India has become a significant center for electronic waste recycling, with an estimated total of 0.8 million tonnes generated in 2012 from domestic and imported sources (Manufacturers' Association for Information Technology, 2013). After unsuccessfully banning imports of e-wastes in the

1990s, the Indian government has more recently introduced regulations aimed at improving recycling (Shinkuma and Managi, 2010). The E-Waste (Management & Handling) Rules were introduced by the government in 2011. e-Waste recyclers must be licensed, but there is a significant informal sector that goes unregulated. Based in cities such as Delhi, this unregulated activity creates serious environmental and health impacts. Concerns also exist over the effectiveness of the official license system (Shinkuma and Managi, 2010).

4.5 CHINA

In terms of the global solar industry, China is of increasing significance as a manufacturer of solar PV technology and the rapid growth of its installed capacity in the past decade. By 2015, capacity had grown to 60 GW and is predicted to rise to 70 GW by 2020 (Deutsche Bank, 2015). According to Yue et al. (2014, p. 670) the majority of manufacturing of PV modules now occurs in Asia, with China the number one country. Production in 2000 was only 1% of the global total but by 2008 this had increased to 33% (Rigter and Vidican, 2010). Yet, despite long established and recently strengthened national environmental regulations, there are significant concerns over the sustainability of the sector in China.

4.5.1 Sector Development Policies—National and Provincial Levels

The massive growth in solar PV installed capacity in China is quite recent. Solar water heating was employed in China during the 1970s (Han et al., 2010), but little PV power was produced until the 1990s. An initial national policy dates back to between 1996 and 2006, when the central government funded and constructed solar PV systems in rural areas through several initiatives (Huo and Zhang, 2012). Sectoral growth was slow during this period, but grew rapidly after the national Renewable Energy Law (see Cherni and Kentish, 2007) entered into force in 2006. It was followed by the *Medium and Long-Term Plan for Renewable Energy*, that set a target of 300 MW of installed solar capacity by 2010 (Yuan et al., 2011). Subsequent policy has sought to continue this expansion, with multiple policy instruments employed.

National policy instruments aimed at increasing installation include: regulations; the building-integrated photovoltaics (BIPV) or building demonstration program; the Solar Roofs Program and the Golden Sun Program, which provide subsidies for small PV projects; in addition to FITs for utility-scale PV projects (Table 4.4). The Renewable Energy Law 2006 compels energy companies to purchase all renewable energy power connected to the grid system and provide grid connections (Huo and Zhang, 2012). Revisions to the legislation in 2009 sought to resolve issues that subsequently emerged from implementation, including a lack of grid connections. China's first major subsidy for solar

TABLE 4.4 Main National Level Policy Instruments for Supporting Solar PV Installation in China

Instrument	Instrument Type	Year of Adoption	Main Aims
Renewable Energy Law	Regulation	2006	To compel energy companies to purchase electricity from renewable sources and provide grid connections
Building-integrated photovoltaics (BIPV) or PV building demonstration program	MBI/subsidy	2009	To subsidize the installation of roof-mounted or wall-mounted solar systems
Golden Sun Program	MBI/subsidy, scientific/ technological support	2009	To support the growth of the PV sector
Feed-in tariffs	MBI	2011	To provide market-based incentives for solar generation

PVs was the BPIV introduced in 2009. Under the scheme, operators were provided with a subsidy per watt of energy produced by wall-mounted or rooftop systems. Under the Golden Sun Program the government aims at using a combination of measures to support growth of the solar PV industry. The 2011 FIT provided a financial incentive for solar generation set at 1 RMB per kWh for projects approved after 2011, and 1.15 RMB for those approved before this date. Recent amendments to this system have updated the FITs to differentiate between different regions in China. Provincial governments have also introduced MBIs, such as FITs. Both Jiangsu and Hebei provinces, for example, offer a FIT for solar based paid for each kilowatt-hour of power produced (Deutsche Bank, 2015).

During 11th Five-Year Plan (2006–10), the Chinese government specifically promoted the solar PV manufacturing industry, leading to its rapid expansion. Several national and provincial policy instruments supported research and development, along with investment in PV production (see Huo and Zhang, 2012, pp. 42–43). According to the government, solar cell production then expanded with an annual growth rate of over 100% (NDRC, 2011, p. 1). By 2010, China had become the number one producer globally, with a market share of 50% (NDRC, 2011). Production was primarily crystalline silicon cells, in which China now dominates production worldwide. Rapid expansion continued thereafter, supported by the 12th Five-Year Plan for the Solar PV industry, issued in 2010. Covering the period between 2011 and 2015, it reviewed the preceding plan before establishing policy goals and measures for this period. In seeking to expand production further, the Plan takes a regulatory, top-down approach in specifying actions required in centralized planning and its implementation by government and industry. Some MBIs are included in the Plan, most notably in the form of subsidies and financial support. By 2013, China accounted for 62% of global production (Yue et al., 2014, p. 670).

4.5.2 Solar PV Production Impact Policies—National and Provincial Levels

This rapid growth of the solar PV manufacturing sector in China has been linked to lower environmental and social production standards than in other countries. Yue et al. (2014) conduct a Life Cycle Assessment of the manufacture of silicon-based solar PV modules in China and Europe. Their findings show that "modules manufactured in China consume 28–48% more primary energy sources," while greenhouse gas emissions from their manufacture were twice those made in Europe (Yue et al., 2014, p. 676). One conclusion made by these researchers is that the lower costs of production in China do not include such environmental externalities. Greenpeace (2013) highlights a litany of additional problems with this industrial sector, particularly toxic waste from production. They argue that only 20 polysilicon manufacturers nationally met Ministry of Industry and Information Technology entry requirements on sustainability for this sector. Solar cell manufacturers are also criticized for not publishing their environmental impacts in CSR reports, with few companies seemingly subject to pollution controls. These issues are also highlighted by Nath (2010), who argues that the growth in silicon-based PV production, particularly in China, raises concerns over toxic wastes such as silicon tetrachloride. He suggests that while manufacturers in the United States and other developed countries have taken steps to recycle this chemical back into production cycles, Chinese companies have ignored this problem to keep costs down.

Resolving environmental impacts from this industry could then involve tightening enforcement. In this respect, China has had a relatively strong legal framework for environmental protection by international comparison for some time, although a consistent problem has been poor implementation by local officials. Demands for economic development have often overridden environmental quality, resulting in serious pollution issues such as contamination of surface and groundwater, and declining urban air quality. Environmental impacts from industrial production are regulated under the national Environmental Protection Law (EPL), originally adopted in 1989 but significantly amended in 2014. The new Law is, according to Zhang and Cao (2015, p. 433), China's "first attempt to harmonize economic and social development with environmental protection." New features introduced include more stringent penalties for infringements of legal requirements, a greater role for public participation, and more provisions for pollution prevention. Yet, there are already concerns over how well these regulations will be implemented: the EPL can be overridden by other policy and ministerial demands; it lacks citizen's rights of engagement; enforcement is weak and decentralized; and China's environmental governance is fragmented within central government (Zhang and Cao, 2015). Local level enforcement is a particular constraint as officials lack authority to penalize polluters, while local governments prioritize economic growth.

4.5.3 Solar PV Installation and End-of-Life Policies—National and Provincial Levels

Theoretically, China has a strong centralized legal framework for governing the impacts of installing and disposing of solar PV systems. All solar projects require authorization from the central government National Development and Reform Commission or provincial equivalent. In addition, approval must be given from several other ministries, including the Ministry of Land and Resources (MLR), the Ministry of Environmental Protection (MEP), and Ministry of Water Resources. A preliminary opinion must be sought from the NDRC, followed by gaining of permission from the relevant ministries. The NDRC will then need to provide final approval. One important consideration for solar project development is an EIA, which is considered by the MEP during the approval process. EIA is widely conducted in China but there are issues with effectiveness.

A compulsion to conduct assessments was introduced by the EPL 1979. Developers were required to submit an EIA of new projects to the relevant environmental protection authorities for planning approval. Subsequent amendments have significantly tightened this compulsion, although a number of studies have identified effectiveness problems with the EIA system (eg, Zhao, 2009). Zhao (2009, p. 493) identifies "serious statutory flaws including insufficient coverage of... projects... excessive power of the local authorities in EIA approval, limited public participation, and lack of deterrence." Large-scale solar projects are subject to an EIA as part of the approval process, but questions still remain over the effectiveness of this process in mitigating environmental impacts.

Toxic waste from electronics is already a critical problem in China. The country is the largest importer of WEEE globally, but standards for recycling and end-of-life disposal of e-waste are often informal, primitive, and lack appropriate health and environmental controls (Yu et al., 2010, p. 991). On paper, China possesses a strong regulatory framework for controlling such impacts. Following a wave of e-waste imports, the government banned informal recycling, but was then faced with significant implementation problems (Shinkuma and Managi, 2010). After this failure, the government has since focused on encouraging a properly licensed and regulated recycling sector, while banning unauthorized activities. As in India, however, this approach relies on enforcement and the imposition of recycler responsibilities, both of which are questionable in this context.

4.6 CONCLUSIONS

By comparing solar PV governance structures between these national contexts several trends are apparent. Firstly, in all contexts, governments have invested significant policy effort in supporting the development of both solar PV production (modules and solar cells) and installed capacity, often with significant results. A variety of policy instruments have been adopted to achieve sectoral growth, primarily regulation, and MBIs. In the United States, governments at federal and state level have provided strong incentives through RESs or RPS, in addition to FITs. This regulatory-FIT approach is visible in Germany, the United Kingdom, and India, with a variety of MBIs employed to provide additional support. India's solar production has however stalled, despite government policy support. China meanwhile has utilized central planning to rapidly expand its solar production and power industry to a position of global leadership within only a decade. While strongly regulatory, China's policy has also introduced market-based innovations.

Secondly, there is significant variation in how production impacts such as water, air, and waste issues, identified in the previous chapter, are governed. These impacts are strictly controlled for production conducted within the United States and EU, even though the majority of solar PV module and cell production now occurs outside of these contexts. US federal legislation provides a strong framework for pollution control in states, while EU environmental policies, in particular IPPC, effectively regulate such effects. In contrast, India has weakly administered environmental regulations, although impacts are limited from its nascent manufacturing industry. Significant problems exist with China's rapidly expanding solar PV sector. In the space of under a decade, this industry has grown from almost zero global market share to market leader. Expansion has come at a significant price in environmental and social externalities, which have not as yet been internalized through effective regulation. Recent amendments to national environmental legislation aim at improving implementation but impacts are likely to continue.

Finally, most countries are attempting to counter impacts from solar PV installation and end-of-life stages. Land use planning systems in Germany and the United Kingdom implement EU legal requirements for EIA, which compel authorities, planners, and developers to consider effects of solar developments on inter alia landscapes, ecology, and communities. Federal agencies and many states in the United States impose similar requirements on projects. Meanwhile, in India and China, EIA is often required but implementation is variable. A similar pattern emerges with regards to waste disposal, where tight regulations exist in the United States and the EU for recycling or landfilling of redundant solar equipment. However, in India and China, such approaches are less developed, suggesting that future environmental problems may emerge as current installations reach the end of their operating capabilities and require replacement.

As our analysis suggests, this variability in governance structures between countries provides a strong rationale for greater international norms and regulatory frameworks. The solar industry is now global in scale, although manufacturing and installation are driven by national policy. This sector is also highly interconnected, with manufacturing shifting to lower production cost countries in East Asia, but high levels of installation in developed countries. In this respect, high environmental quality in developed countries may be offset to developing countries, where there appears to be a race to the bottom in both environmental and social standards. In the next chapter, we will therefore explore the potential for adopting an international governance framework to mitigate environmental and social impacts from this industrial sector and ensure its long-term sustainability.

ABBREVIATIONS

AATF	Approved Authorised Treatment Facility
AEs	Approved Exporters
BAT	Best Available Techniques
BATRRT	Best Available Treatment, Recovery and Recycling Techniques
BauGB	Federal Building Code (Germany)
BFRs	brominated flame retardants
BIPV	building-integrated photovoltaics
BREFs	BAT Reference guidance documents
CEQA	California Environmental Quality Act
CfD	contracts for difference
CTE	consents to establish
CTO	consent to operate
DCR	domestic content requirement
DTSC	Department of Toxic Substances Control
EEC	European Economic Community
EIA	environmental impact assessment
EIB	European Investment Bank
EICC	Electronic Industry Citizenship Coalition
EIR	environmental impact report
ElektroG	Elektro- und Elektronikgeurategesetz (Germany)
ELVs	Emission Limit Values
EMAS	Eco Management and Audit System
EPA	Environmental Protection Agency
EPL	Environmental Protection Law

EPR	Environmental Permitting (England and Wales) Regulations
EU	European Union
FIT	feed-in tariff
GBI	generation-based incentive
GW	giga watts
HWCL	Hazardous Waste Control Law
IED	Industrial Emissions Directive
IPPC	integrated pollution and prevention control
IREDA	Indian Renewable Energy Development Agency Ltd
ITC	Investment Tax Credit
JNNSM	Jawaharlal Nehru National Solar Mission
KrWG	Waste Management Act (Germany)
MACRS	Modified Accelerated Cost Recovery System
MBIs	market-based instruments
MEP	Ministry of Environmental Protection
MLR	Ministry of Land and Resources
MNRE	Ministry of New and Renewable Energy
MSDS	Materials Safety Data Sheet
MW	mega watt
NEPA	National Environmental Policy Act
NPPF	National Planning Policy Framework
NTP	National Tariff Policy
PBDEs	processing of brominated diphenylethers
PCS	Producer Compliance Scheme
PPAs	Power Purchase Agreements
PURPA	Public Utility Regulatory Policy Act
PV	photovoltaics
RCRA	Resource Conservation and Recovery Act
REA	Renewable Energy Act
RES	Renewable Energy Standard
RHI	Renewable Heat Incentive
RO	Renewables Obligation
ROCs	Renewable Obligation Certificates
RoHS	reduction in the use of hazardous substances
RPO	Renewable Purchase Obligation
RPS	Renewable Portfolio Standards
SECI	Solar Energy Corporation of India
SEIA	The Solar Energy Industries Association
SPCB	State Pollution Control Board
SERCs	State Electricity Regulatory Commissions
SRECs	State Renewable Energy Certificates
SVTC	Silicon Valley Toxics Coalition
TCLP	Toxicity Characteristic Leach Procedure
TREC	tradable renewable energy credit

TRI	Toxics Release Inventory
USA	United States of America
USEPA	US Environmental Protection Agency
WEEE	waste electrical and electronic equipment

REFERENCES

Andrews, R.N.L., 2006. Managing the Environment, Managing Ourselves: A History of American Environmental Policy. Yale University Press, New Haven.

Benson, D., Jordan, A., 2011. What have we learned from policy transfer research? Dolowitz and Marsh revisited. Polit. Stud. Rev. 9 (3), 366–378.

Benson, D., Jordan, A., 2013. Environmental policy. In: Cini, M., Pérez-Solórzano Borragán, N. (Eds.), European Union Politics. fifth ed. Oxford University Press, Oxford.

Benson, D., Jordan, A., 2016. Climate policy instrument choices. In: Farber, D., Peeters, M. (Eds.), Climate Change Law. Edward Elgar, Cheltenham.

Building Research Establishment (BRE), 2013. Planning Guidance for the Development of Ground Mounted Solar PV Systems. BRE, Watford.

California Public Utilities Commission (CPUC), 2015. Report to the Legislature in Compliance with Public Utilities Code Section 910. CPUC, San Francisco.

Cherni, J.A., Kentish, J., 2007. Renewable energy policy and electricity market reforms in China. Energy Pol. 35, 3616–3629.

Connelly, J., Smith, G., Benson, D., Saunders, C., 2012. Politics and the Environment: From Theory to Practice. Routledge, London.

Deutsche Bank, 2015. Crossing the Chasm. Deutsche Bank AG, Frankfurt, Germany.

DSIRE (Database of State Incentives for Renewables and Efficiency), 2015. Renewable Portfolio Standard Policies. http://www.dsire.org/rpsdata/index.cfm (accessed 06.03.15.).

Environmental Protection Agency, 2012. TCLP Questions. USEPA, Washington. http://www.epa.gov/epawaste/hazard/testmethods/faq/faq_tclp.htm. (accessed 06.05.15.).

Environmental Protection Agency, 2015. Summary of the Resource Conservation and Recovery Act. USEPA, Washington. http://www2.epa.gov/laws-regulations/summary-resource-conservation-and-recovery-act. (accessed 06.05.15.).

European Photovoltaic Industry Association (EPIA), 2014. Global Market Outlook for Photovoltaics 2014–2018. EPIA, Brussels.

Fischman, R., 2005. Cooperative federalism and natural resources law. NYU Environ. Law J. 14, 179.

Fraunhofer Institute, 2015. Recent Facts About Photovoltaics in Germany. Fraunhofer Institute for Solar Energy Systems, Freiburg.

Government of Andhra Pradesh, 2015. Andhra Pradesh Solar Power Policy, 2015. Government of Andhra Pradesh, Hyderabad.

Government of Karnataka, 2014. Solar Policy 2014–2021. Government of Karnataka, Bangalore.

Government of Rajasthan, 2014. Draft Rajasthan Solar Energy Policy, 2014. Government of Rajasthan, Jaipur.

Government of Tamil Nadu, 2012. Tamil Nadu Solar Energy Policy. Government of Tamil Nadu, Chennai.

Greenpeace, 2013. Unravelling the Puzzle That is Solar PV Pollution—Clean Production of Solar PV Manufacture in China. Greenpeace, Amsterdam.

Han, J., Mol, A.P.J., Lu, Y., 2010. Solar water heaters in China: a new day dawning. Energy Pol. 38, 383–391.

Huo, M.-L., Zhang, D.-W., 2012. Lessons from photovoltaic policies in China for future development. Energy Pol. 51, 38–45.

Jordan, A., Benson, D., Wurzel, R.K.W., Zito, A.R., 2012a. Environmental policy: governing by multiple policy instruments? In: Richardson, J. (Ed.), Constructing a Policy-Making State? Policy Dynamics in the European Union. Oxford University Press, Oxford.

Jordan, A., Benson, D., Wurzel, R.K.W., Zito, A.R., 2012b. Governing with multiple policy instruments? In: Jordan, A.J., Adelle, C. (Eds.), Environmental Policymaking in the European Union. Earthscan, London.

Jordan, A., Wurzel, R.K.W., Zito, A.R., 2013. Still the century of 'new' environmental policy instruments? Exploring patterns of innovation and continuity. Environ. Polit. 22 (1), 155–173.

Khare, V., Nema, S., Baredar, P., 2013. Status of solar wind renewable energy in India. Renew. Sustain. Energy Rev. 27, 1–10.

Kjaer, A.M., 2005. Governance. Polity Press, Cambridge.

Lorenzoni, I., Benson, D., 2014. Radical institutional change in environmental governance: explaining the origins of the UK Climate Change Act 2008 through discursive and policy streams perspectives. Glob. Environ. Chang. 29, 10–21.

Majone, G., 1994. The rise of the regulatory state. West Eur. Polit. 17 (3), 77–101.

Manufacturers' Association for Information Technology (MAIT), 2013. What is E-Waste. http://mait.com/ewaste/about.html (accessed 15.05.15.).

Ministry of New and Renewable Energy (MNRE), 2013. Developmental Impacts and Sustainable Governance Aspects of Renewable Energy Projects. Government of India, New Delhi.

Murthy, V.A.V., 2014. India's Solar Energy Future: Policy Institutions. Center for Strategic and International Studies, Washington, DC.

Nath, I., 2010. Cleaning up after clean energy: hazardous waste in the solar industry. Stan. J. Int. Relat. XI (2), 6–13.

National Development and Reform Commission (NDRC), 2011. 12th Five-Year Plan for the Solar Photovoltaic Industry. NDRC, Beijing.

Official Journal, 2011. Directive 2011/65/EU of the European Parliament and of the Council of the 8 June 2011 on the Restriction of the Use of Certain Hazardous Substances in Electrical and Electronic Equipment. L 174/88. Official Journal of the European Communities, Luxembourg.

Panigrahi, J.K., Amirapu, S., 2012. An assessment of EIA system in India. Environ. Impact Assess. Rev. 35, 23–56.

Pierre, J., Peters, B.G., 2000. Governance, Politics and the State. Palgrave Macmillan, Basingstoke.

Rigter, J., Vidican, G., 2010. Cost and optimal feed-in tariff for small scale photovoltaic systems in China. Energy Pol. 38, 6989–7000.

Russel, D., Benson, D., 2014. Green budgeting in an age of austerity: a transatlantic comparative perspective. Environ. Polit. 23 (2), 243–263.

Sahoo, A., Shrimali, G., 2013. The effectiveness of domestic content criteria in India's Solar Mission. Energy Pol. 62, 1470–1480.

Sharma, N.K., Tiwari, P.K., Sood, Y.R., 2012. Solar energy in India: strategies, policies, perspectives and future potential. Renew. Sustain. Energy Rev. 16, 933–941.

Shinkuma, T., Managi, S., 2010. On the effectiveness of a license scheme for E-waste recycling: the challenge of China and India. Environ. Impact Assess. Rev. 30, 262–267.

Solar Energy Industries Association (SEIA), 2015a. Federal Clean Energy Contracting. SEIA, Washington. http://seia.org/policy/renewable-energy-deployment/federal-clean-energy-contracting (accessed 06.05.15.).

Solar Energy Industries Association (SEIA), 2015b. Renewable Energy Standards. SEIA, Washington. http://seia.org/policy/renewable-energy-deployment/renewable-energy-standards (accessed 06.05.15.).

Solar Energy Industries Association (SEIA), 2015c. Solar Tax Exemptions. SEIA, Washington. http://seia.org/policy/renewable-energy-deployment/solar-tax-exemptions(accessed 06.05.15.).

US Treasury, 2015. 1603 Program: Payments for Specified Energy Property in Lieu of Tax Credits. Department of the Treasury, Washington. http://www.treasury.gov/initiatives/recovery/Pages/1603.aspx (accessed 06.05.15.).

UK Department of Business, Innovation and Skills, 2013. WEEE Regulations 2013: Government Guidance Notes. UK Government, London.

Ummadisingu, A., Soni, M.S., 2011. Concentrating solar power—technology, potential and policy in India. Renew. Sustain. Energy Rev. 15, 5169–5175.

US Department of Energy, 2015a. Modified Accelerated Cost-Recovery System (MACRS) + Bonus Depreciation (2008–2012). Department of Energy, Washington. http://energy.gov/savings/business-energy-investment-tax-credit-itc (accessed 06.05.15.).

US Department of Energy, 2015b. Powering New Markets: Utility-Scale Photovoltaic Sector. Department of Energy, Washington.http://energy.gov/sites/prod/files/2015/02/f19/DOE_LPO_Utility-Scale_PV_Solar_Markets_February2015.pdf.

US Department of Energy, 2015c. Business Energy Investment Tax Credit (ITC). Department of Energy, Washington. http://energy.gov/savings/business-energy-investment-tax-credit-itc (accessed 06.05.15.).

Watts, R.L., 2008. Comparing Federal Systems. Queen's University, Kingston, ON.

Wurzel, R.K.W., Zito, A.R., Jordan, A.J., 2013. Environmental Governance in Europe: A Comparative Analysis of New Environmental Policy Instruments. Edward Elgar, Cheltenham.

Yu, J., Williams, E., Ju, M., Shao, C., 2010. Managing e-waste in China: policies, pilot projects and alternative approaches. Resour. Conserv. Recycl. 54, 991–999.

Yuan, X., Zuo, J., Ma, C., 2011. Social acceptance of solar energy technologies in China—end users' perspective. Energy Pol. 39 (3), 1031–1036.

Yue, D., You, F., Darling, S.B., 2014. Domestic and overseas manufacturing scenarios of silicon-based photovoltaics: life cycle energy and environmental comparative analysis. Sol. Energy 105, 669–678.

Zhang, B., Cao, C., 2015. Policy: four gaps in China's new environmental law. Nature. 21 January, http://www.nature.com/news/policy-four-gaps-in-china-s-new-environmental-law-1.16736.

Zhao, Y., 2009. Assessing the environmental impact of projects: a critique of the EIA legal regime in China. Nat. Resour. J. 49, 485–524.

Chapter 5

A Normative Perspective on Governing Solar PV Sustainability

5.1 INTRODUCTION

As outlined in the previous chapters, solar photovoltaic (PV) technologies are rapidly expanding on a global scale, but have multiple potential impacts with implications for governance. Chapter 3 primarily focused on health and toxicological impacts, but other effects are associated with this sector, including those from production processes such as greenhouse gas emissions (GHG), in addition to the sociocultural effects on landscapes caused by the installation of solar arrays. As Chapter 4 shows, multiple policy instruments have been employed to both support the development of solar PV production, installation and power generation, and counter some of these impacts, although responses to the latter have proved variable with many regulatory gaps apparent. While these chapters have therefore helped empirically map the emerging landscape of solar PV governance, critical questions remain over the extent to which sustainability is being governed and, in a normative sense, how it should be governed in the future as part of a global green economy (see Barbier and Markandya, 2013).

Our analysis of governance instruments in different countries reveals several patterns. Most notably, national approaches to regulating impacts vary significantly in terms of instruments adopted. In Western economies such as the United Kingdom and Germany, a dense regulatory framework of existing measures—much of it derived from EU legislation—should counter most environmental and social risks from expansion of this sector. In the United States, environmental impacts are also strongly regulated by federal and state legislation and policy. China and India, however, have adopted a less regulatory approach, instead concentrating on promoting sector expansion via national or regional policy incentives. Far less attention has been paid by governments to controlling environmental and social impacts from production, operation, and disposal of redundant solar PV systems. This unequal pattern highlights a paradox as manufacturing shifts from countries such as the United States to India, China, and other newly industrializing states, implying a need for a more uniform approach to governance between countries. Without such

Solar Photovoltaic Technology Production. http://dx.doi.org/10.1016/B978-0-12-802953-4.00005-6

harmonization, what economists call a "race to the bottom" (see Revesz, 1992) can occur in which expanding solar PV energy demand in Western states offsets its production-related environmental externalities to developing countries which then compete on lower standards of social and environmental quality. Offsetting of production externalities in this way could then create so-called pollution havens (Cole, 2004) through the globalization of such trade. Countering these externalities through harmonization of governance, however, remains potentially problematic on a global scale due to the nature of the international political-economic system.

Two potential constraints are apparent. Firstly, intensifying globalization of economic activity means that governing sustainability issues has become increasingly complex (Connelly et al., 2012). Commentators have argued that globalization has undermined the sovereign ability of states to govern global economic forces and its external effects (eg, Strange, 1996; Weiss, 1999[1]). The solar PV sector increasingly operates on a transboundary scale, with consumption of technologies becoming separated transnationally from its production and hence sustainability impacts. This pattern of economic development is reflective of the arguments made by supporters of globalization, who maintain that it allows for a more efficient spatial allocation of productive activity (eg, Massey, 1995), but clearly these arguments do not readily account for the environmental or social race to the bottom in their efficiency calculations. Secondly, the capacity of states to cooperate in response to such threats is impaired by what some international relations scholars call the "anarchy" of the global political system, whereby no credible authority exists above the level of the nation state to impose cooperative solutions. According to neorealist IR authors, anarchical relations can, to an extent, be overcome by the creation of international regimes[2] or agreements, where powerful hegemonic states force cooperation to maximize their own gains, but they also argue that such compacts are unstable or ephemeral (eg, Hasenclever et al., 1997). Global climate change negotiations in Copenhagen 2009, under the auspices of the United Nations Framework Convention on Climate Change (UNFCCC), ably demonstrate this cooperation dilemma (Connelly et al., 2012). Nonetheless, states can and indeed do cooperate successfully to frame effective environmental regimes, for example, the UNFCCC Paris Conference 2015, demonstrating degrees of harmonization are possible where regimes can be agreed. Several hundred international environmental regimes have to date been adopted (see Sands and Peel, 2012), covering a diversity of issues from biodiversity loss to marine pollution, suggesting global responses to solar PV impacts are theoretically possible.

---

1. For counterarguments regarding the capacity of globalization to weaken state power, see Pierre (2013).
2. Krasner (1983, p. 1) refers to regimes as comprising "principles, norms, rules, and decision-making procedures around which actor expectations converge in a given issue-area."

In addition, both state and nonstate actors, such as business could do more to counter these threats themselves.

With these constraints in mind, this chapter explores the potential for more effectively governing solar PV impacts globally. As it goes on to discuss in the next section, by drawing on existing guidelines and academic arguments to frame a discussion, some options may be viable. Rather than a strong, legally binding agreement, which under conditions of global anarchy may prove infeasible to fashion, one approach could involve developing a set of normative global principles for the solar PV sector to guide its sustainability in all countries. National governments could employ them to consider greater regulation of impacts, using different policy instruments, such as controls on emissions or environmental impact assessments (EIAs) of solar projects. Given the transnational nature of the solar PV sector, these principles could also be used to further promote industry-led voluntary approaches that utilize existing governance mechanisms such as corporate social responsibility (CSR), environmental management systems (EMS), life cycle assessment (LCA) of products, and extended producer responsibility (EPR). Significant scope, we consequently argue, exists for using multi-level governance and governance "beyond the state" to better enhance the sustainability of the sector and contribute to wider global norms such as the UN's Sustainable Development Goals (SDGs).

5.2 SOLAR PV GOVERNANCE—SUSTAINABILITY PRINCIPLES IN PRACTICE

A critical question for examination is how, given the uneven geo-political landscape of governance responses described in Chapter 4, should the sustainability of solar PVs be governed? The notion of sustainability incorporates the principle of sustainable development, defined as "[d]evelopment that meets the needs of the present without compromising the ability of future generations to meet their own needs" (WCED, 1987, p. 43). Achieving this aim has subsequently been interpreted to mean integrating environmental, economic, and social objectives within development to meet such needs. Yet, due to the anarchic relations of the international order, a strong, legally binding approach to embedding sectoral sustainability in solar PV governance would perhaps be unrealistic. Previous international-level responses to sustainability have largely relied upon agreements between states, most notably the 1992 Rio United Nations Conference on Environment and Development (UNCED) and its main outputs, the Rio Declaration and Agenda 21 (Baker, 2006). While national governments agreed to implement these regimes through their domestic policies and sustainable development strategies, implementation has been poor and uncoordinated (see Lafferty and Meadowcroft, 2000; Connelly et al., 2012). Consequently, one output of the follow-up Rio +10 Conference in Johannesburg 2002 was determining an increasing role for nonstate actors in governing sustainability through so-called partnership type approaches, with the UN taking a steering capacity (Connelly et al., 2012). Further embedding of sustainability norms into global governance has occurred through the development of the 2000 Millennium Development Goals (MDGs) and, more recently, the follow-up SDGs adopted in 2015 (UN, 2015). As discussed below, the SDGs will form an over-arching normative framework to guide governments and private actors in shaping sustainable development worldwide up to 2030 (UN, 2015). Governing the global sustainability of the solar PV sector should, it is argued, therefore seek to better integrate its future development with this more normative and nonbinding approach, as this would perhaps prove more acceptable to state sovereignty concerns and the interests of multinational companies. But on a global scale, few normative guidelines to ensure the sustainability of the solar industry are evident. Where guidance does exist, it is largely national or company specific, patchy, and uncoordinated. This section therefore reviews this, albeit limited, guidance and reflects back on analyses of solar PV impacts to help draw out some potential principles for a harmonized global response.

One widely known example of sustainability guidance comes from the United States, where the Solar Energy Industry Association (SEIA) has established a Solar Industry Environmental & Social Commitment. The SEIA is a trade organization which represents the interests of industries that manufacture, install, and operate solar PV technologies in the United States. A central component of its approach to embedding industry-wide sustainability is the Commitment, to which members can sign up, although participation is not mandatory (SEIA, 2012). Participants must submit a letter of intent or sign a declaration of support to be included in the Commitment. Key performance indicators (KPIs) must then be met, covering 13 areas such as water consumption and waste generation. These KPIs are used as a basis for annual benchmarking and reporting. Companies must also show that they meet the SEIA criteria for implementation and education, reporting and transparency, continuous improvement, and future development (SEIA, 2012). Despite a requirement to report annually, the Commitment is voluntary and industries are encouraged to evaluate progress and continuously improve sustainability performance.

Staying in the United States, the Silicon Valley Toxics Coalition (SVTC), a nonprofit organization dedicated to improving the environmental performance of high-tech manufacturing, has developed several initiatives aimed at the sustainability of the solar PV sector. In its 2009 White Paper (SVTC, 2009), the Coalition identifies specific threats associated with solar PV technologies from hazardous materials. Here, it makes recommendations for a "Clean and Just Solar Industry" that include: reducing and eliminating toxic material usage; greater industry accountability; effective testing of technologies; recycling of PVs; promoting better employment; and, protecting the health and safety of consumers (SVTC, 2009). To support its aims, the SVTC has developed a solar "scorecard" that members can employ to check the sustainability performance of solar PV modules, with results published annually by the Coalition (SVTC, 2015). The scorecard operates as a weighted matrix that awards high scores for supporting EPR (discussed later), upholding worker rights, ensuring health and safety plus the sustainability of supply chains, emissions transparency, and reducing module toxicity. Other points are awarded for chemical use reduction, recycling, avoiding biodiversity impacts, energy efficiency, reductions in water consumption, nonutilization of prison labor and avoidance of conflict minerals. Companies then receive an overall sustainability rating, scored out of 100.

Although it does not publish performance criteria, the European Photovoltaic Industry Association provides some thematically based guidance "fact sheets" on sustainability issues, limited to materials availability, external costs, value chains, job creation, water use, land use and biodiversity, carbon footprints, and energy payback time (EPBT) (EPIA, 2015a). Water footprints, for example, are discussed for PV systems throughout their lifecycles. As identified, while water consumption is decoupled from PV electricity generation, providing positive impacts, manufacturing and recycling require water inputs (EPIA, 2015b). As water consumption of PVs is calculated at 0.1 L/kWh, compared to between 0.75 and 75 L/kWh for conventional fossil fuel generation, solar PVs are argued to provide positive benefits for sustainable development (EPIA, 2015b). Similar positive impacts are claimed for EPBT, biodiversity, carbon footprints, and other sustainability aspects. Again, the approach adopted is largely industry-led and reliant on information provision by companies.

Other national and international associations issue sustainability guidance to their membership, although approaches are far from unified. The Electronic Industry Citizenship Coalition (EICC) represents a number of leading, mainly US, companies with the aim of creating industry standards for ethical, social, and environmental aspects of electronic industry supply chains. To this effect, the EICC issues its own Code of Conduct for members, with the latest version adopted in 2014 (EICC, 2014). The Code draws on international norms for human rights, labor standards, and environmental and ethical practice, with the aim of ensuring that "working conditions in the electronics industry supply chain are safe, that workers are treated with respect and dignity, and that business operations are environmentally responsible and conducted ethically" (EICC, 2014, p. 1).

In addition, many leading solar PV companies implement industry-specific voluntary initiatives for embedding aspects of sustainability into their business models: CSR, LCAs, EMS, and EPR. As discussed below, these voluntary mechanisms could support the integration of dedicated sustainability principles across the sector. EPR also offers specific advantages for the governance of toxicological and waste impacts from end-of-life products.

5.3 NORMATIVE SUSTAINABILITY PRINCIPLES FOR GLOBAL SOLAR PV GOVERNANCE?

Some universal principles can be drawn out from these industry standards and impacts identified in the scientific literature to guide global governance. These principles could relate to specific sustainability themes that are evident within solar PV production, installation and disposal, namely: toxicology and human health; energy efficiency; labor; waste generation and disposal; climate emissions; ecosystems; and conflict minerals. Their implementation can be measured through related sustainability KPIs, which in turn integrate with broader, strategic level policy in the form of the UN's SDGs. Such principles can be utilized at multiple governance levels and by multiple actors, most notably governments in regulating (or supporting) the solar PV sector through sectoral policy and specific instruments such as EIA and—more significantly—business via preexisting instruments such as CSR, EMS, LCA, and EPR.

5.3.1 Sustainability Principles

The eight sustainability themes and principles were developed specifically to encompass the positive and negative externalities of solar PV technologies. Toxicological and health impacts are discussed in depth in Chapter 3, but development of these principles also draws on existing industry sustainability standards identified above and the scientific literature. To fully address sustainability, the principles cover environmental, social, and economic aspects (Table 5.1). They also link to a series of associated KPIs, designed to measure their application in practice, and the UN's SDGs, discussed further below.

TABLE 5.1 Normative Sustainability Principles for Global PV Governance

Sustainability Themes	Sustainability Principles	KPIs	SDGs
Toxicology and health	To protect human health throughout PV life cycles	Toxicity of materials in the mining, manufacture, and end-of-life disposal phases	*Goal 3:* Ensure healthy lives and promote well-being for all at all ages
		Health of workers	
		Levels of disease linked to air pollutants	
Energy efficiency	To reduce energy use throughout PV life cycles and enhance the efficiency of PV systems	Energy use and generation Energy payback time (EPBT)	*Goal 7:* Ensure access to affordable, reliable, sustainable, and modern energy for all

Continued

TABLE 5.1 Normative Sustainability Principles for Global PV Governance—cont'd

Sustainability Themes	Sustainability Principles	KPIs	SDGs
Labor	To protect the employment rights of workers throughout PV life cycles	International Labour Organization standards and international human rights norms	Goal 8: Promote sustained, inclusive, and sustainable economic growth, full and productive employment and decent work for all
Waste reduction and disposal	To reduce waste throughout solar PV life cycles	Levels of waste production Levels of recycling	Goal 12: Ensure sustainable consumption and production patterns
Climate	To reduce GHG emissions throughout solar PV life cycles	GHG emissions from solar PV life cycles	Goal 13: Take urgent action to combat climate change and its impacts
Ecosystems	To reduce impacts on land, water, and biodiversity resources throughout solar PV lifecycles	Land take Water use Toxic emissions to air, water, and land Biodiversity loss Visual impacts	Goal 15: Protect, restore, and promote sustainable use of terrestrial ecosystems
Conflict minerals	To eliminate the use of conflict minerals in solar PV production	Use of conflict minerals	Goal 16: Promote peaceful and inclusive societies

KPIs, key performance indicators; SDGs, sustainable development goals.

A critical production and end-of-life disposal impact identified in Chapter 3 is the *toxicity* of materials employed in some solar PV technologies, with their attendant impacts on *human health*. As discussed, particular threats exist from in the mining of silica and crystalline silicon (c-Si) wafer production, in both its crystalline and amorphous forms. Dust from silica mining can, for example, cause lung disease in miners. Up to 50% of c-Si waste dust, or kerf, is lost to production through emissions to air and water, which also pose health risks to workers. Hazardous materials are also employed in silicon processing (Table 3.1). Thin-film photovoltaics pose additional threats to human health. Cadmium telluride (CdTe) also poses hazards due to the toxicity of this chemical, while copper indium selenide (CIS) and copper indium gallium selenide (CIGS) solar PV technologies also involve the use of toxic materials. Gallium arsenide, employed in some solar cells, contains arsenic which is highly toxic to humans. While such threats are most acute in the production phase, waste disposal also presents potential impacts for recyclers and those in proximity to waste dumps. A critical sustainability principle should consequently be to protect human health from such chemicals throughout solar PV lifecycles, including end-of-life disposal. KPIs of sustainability therefore relate to the toxicity of materials included in solar PV technologies and the health of production workers, but also those responsible for recycling and final disposal. Another KPI could be the associated health benefits of solar PV energy production. Research has shown that serious diseases such as heart attacks and asthma could be ameliorated through reductions in SO_2 and NO_x air pollutants produced by fossil fuel energy generation (Hosenuzzaman et al., 2015). An additional KPI could then relate to societal levels of disease linked to air pollution.

Another significant positive economic externality associated with the solar PV sector is its contribution to clean, renewable, and efficient energy production. *Energy efficiency* of the sector is therefore a critical sustainability theme for both production and operation of solar PV systems. Indicators of energy use could focus on calculations of energy consumption in the different stages of solar cell production, to identify ways of reducing energy inputs. In addition, solar PVs have the potential to significantly enhance the efficiency of energy production vis-a-vis conventional fossil fuel sources, thereby providing economic, environmental, and social benefits. Solar water heating (SWH) technology in Zhejiang, China has reduced fossil fuel use while preventing emissions of SO_2, NO_2, dust, and CO_2 (Han et al., 2010). Social benefits of SWH include economically efficient, safe and clean hot water that also reduces CO poisoning risks (from gas-fired heaters) (Han et al., 2010). For operators of solar PV energy systems, a critical indicator is the EPBT, the time required to "compensate for the total primary energy (renewable and nonrenewable) required throughout the life cycle of an energy supply system" (Yue et al., 2014, p. 672), as it provides an important measure of efficiency. Research shows that EPBT can vary significantly according to solar PV systems. For example, Si-PV modules produced in China are calculated to use 28–48% more energy to produce than those manufactured in Europe (Yue et al., 2014). However, Garcia-Valverde et al. (2009, p. 1445) show that while payback of embodied energy of a specific PV installation in Spain was 9.08 years of operation, emissions of 131 g of CO_2 per kWh were estimated for its 20-year life-span; significantly lower than alternative diesel generation or electricity supplied through the national grid. Measuring energy efficiency of solar PVs could then be achieved through KPIs such as levels of energy used for production, energy generation and EPBT.

Another impact of the global solar PV industry, recognized in the SVTC and EICC codes of practice, is the use of both forced *labor* and poor working conditions of workers. Indeed, a specific criterion is included in the SVTC's scorecard for supporting the nonemployment of prison labor. Significant issues have also emerged over the disposal of waste electronic equipment in countries such as China and India (Shinkuma and Managi, 2010), with child labor and the exploitation of workers. Ghana has also become a center for re-processing EU and US electrical and electronic waste, which is often conducted by workers lacking basic protection from hazardous materials (BBC, 2016). Although individual countries effectively decide on labor employment laws, the conditions for workers in the European Union and the United States are invariably stringent. On a global scale, there is wide variability in the protection of workers generally, but nonetheless international norms and agreements exist for labor standards and human rights. The UN Declaration of Human Rights 1948 provides a broad principled framework, particularly Article 23 which inscribes into international law "the right to work, to free choice of employment, to just and favorable conditions of work and to protection against unemployment" (UN, 1948). These legal rights underpin multiple international treaties, but labor rights are most strongly promoted globally by the International Labour Organization (ILO), a UN agency. Adopted in 1998, the ILO has produced its own declaration on fundamental principles and rights at work that contains four principles relating to freedom of association and collective bargaining, elimination of forced or compulsory labor, abolition of child labor, and elimination of employment or occupational discrimination (ILO, 2015). These international norms and principles could support sustainability principles for the solar PV sector and could be employed as KPIs (Table 5.1). They would also strongly integrate with the UN's SDGs, further discussed below.

End-of-life solar PV panels can present negative and positive impacts in terms of waste disposal. Clearly, if valuable materials can be recovered and potential toxicological effects from land-filling panels avoided, then recycling is optimal. The recycling of silicon and CdTe PV modules is widely happening, with extra care taken to avoid landfilling and reuse of precious materials such as tellurium (Te). The SVTC therefore includes a criterion for "High Value Recycling" in its reporting scorecard, which involves achieving 95% recycling of the module into products of equivalent quality and value, with work conducted in facilities operating EMSs (ISO14001) under high standards of worker safety (SVTC, 2015). But problems can arise from recycling processes. For example, Rocchetti and Beolchini (2015) compare two materials recycling scenarios for end-of-life CdTe and CIGS thin-film photovoltaic panels. Using LCA to measure several indicators (abiotic depletion, acidification potential, eutrophication potential, global warming potential, ozone layer depletion potential, and photochemical ozone creation), they show that despite some positive impacts from the recycling of materials, currently "the disposal of end-of-life panels is not advantageous for the environment" (Rocchetti and Beolchini, 2015). That said, their research identifies innovative, environmentally efficient recycling processes for CdTe panels and presents recommendations on how recycling of PV panels can be made more effective in the future, particularly through the recovery of tellurium, indium, gallium, and selenium. Levels of waste production and recycling could then provide KPIs for these impacts.

Solar PV technologies are widely promoted as "clean" zero emission or emissions-free systems of energy production, so should provide positive impacts for *climate*. Not only do these technologies provide green energy when operational, they also reduce reliance on climate polluting fossil fuel generation, which is particularly significant for emerging economies such as India and China. However, when considering production of PV systems, some studies have identified negative impacts associated with GHGs, due to energy inputs. For example, GHG emissions from Si-PV modules manufactured in China are double those in Europe (Yue et al., 2014, p. 676). KPIs for implementing this principle could then include GHG emissions across solar PV life cycles (Table 5.1).

Chapter 3 also identified a range of potential toxicological impacts with implications for both man-made and natural ecosystems. A critical consideration for sustainability therefore is any potentially negative impacts from production discharges to land, water, and air. Again, while tightly controlled in the EU, via IPPC legislation, and the United States by federal and state regulations, these emissions are often subject to laxer controls in newly developing economies. Water use in solar PV production could also present problems, particularly in water-stressed areas where such industries can compete with other water users, including agriculture and public suppliers. In a positive sense, solar PV power generation requires less water for cooling than fossil fuel plants and hence, may help reduce pressure on these resources. Solar installations can require large amounts of space, so impacts on land and biodiversity can be significant. Other problems experienced in countries such as the United Kingdom include the visual impact of large solar PV installations, meaning that some projects have been refused planning consent. Han et al. (2010, p. 389) describe how SWH technology in China "does not integrate easily into most existing buildings" and while "solar collectors are eye-catching on the roof... they cause visual pollution." Therefore, a principle is included for impacts on land, water, and biodiversity resources. Multiple indicators could support this principle, for example, land resources used for solar arrays, visual impacts of facilities, water used in the production of solar PV panels (or saved from the replacement of fossil fuel energy generation), and associated biodiversity loss. Multiple KPIs for an ecosystems principle could then include water use and toxic emissions in production processes, in addition to land take, biodiversity loss, and visual impacts for solar PV energy production.

Another controversial issue concerning the global solar PV industry and indeed, the wider electronics sector, is the use of *conflict minerals*. The illicit export of minerals such as tin from conflict zones, primarily in central Africa, can support the exploitation of children, slavery, illegal arms trading, and armed insurgencies. The SVTC therefore includes a specific criterion in its scorecard for conflict minerals to ensure that "the company has done due diligence and can confirm with reasonable certainty that they do not use conflict minerals" from several African states identified by OECD guidance (SVTC, 2015). Eliminating the use of such minerals from supply chains therefore severs the links between solar PV production and the continuation of these conflicts while supporting the sustainable development of such countries. Some leading manufacturers have reacted to this problem by developing their own conflict minerals protocols and policy. For example, the SunPower Corporation and its subsidiaries have adopted a Conflict Minerals Policy (SunPower, 2015) that aims to eliminate human trafficking and slavery from direct supply chains. In this respect, the company has committed itself to inter alia: conducting a human trafficking risk assessment of supply chains; undertaking audits of suppliers to ensure their compliance with company standards where required; asking direct material suppliers to sign a contract amendment stating that they are not engaged in such activities; and, demanding suppliers go beyond domestic legal requirements and draw upon international norms such as the UN Declaration of Human Rights and the UN Global Compact when eliminating slavery and human trafficking from their operations (SunPower, 2015). Other leading solar companies are also adopting voluntary responses. For example, First Solar publishes a conflict minerals report showing how it is eliminating them from its production, installation, and maintenance of solar PV modules. This report is published in respect of the legal requirements of Section 1502 of the US federal Dodd-Frank Wall Street Reform and Consumer Protection Act 2010[3] that seek to eliminate the use of conflict minerals and associated human rights abuses (First Solar, 2015).

---

3. Section 1502 of the Act compels the Securities and Exchange Commission to "promulgate rules requiring issuers with conflict minerals that are necessary to the functionality or production of a product manufactured by such a person to disclose annually whether any of those minerals originated in the Democratic Republic of the Congo or an adjoining country" (Securities and Exchange Commission, 2012, p. 1). Consequently, the Act then requires "the issuer to submit a report to the commission that includes a description of the measures it took to exercise due diligence" in sourcing the minerals (Securities and Exchange Commission, 2012, p. 1).

5.3.2 Sustainable Development Goals

The principles and KPIs in Table 5.1 provide strong linkage to the UN's SDGs, formally adopted in 2015. UN actors have agreed a total of 17 SDGs and 169 targets to guide international development up to the year 2030. The SDGs cover a range of economic, social, and environmental objectives from ending poverty to protecting ecosystems (Box 5.1).

Returning to our sustainability principles, the theme for *human health* provides integration with the SDGs in the form of Goal 3, which supports healthy lives and well-being globally. Target 3.9 of this Goal specifically aims to "substantially reduce the number of deaths and illnesses from hazardous chemicals and air, water and soil pollution and contamination" (UN, 2015).

BOX 5.1 The United Nations Sustainable Development Goals (SDGs)

Goal 1: End poverty in all its forms everywhere.

Goal 2: End hunger, achieve food security and improved nutrition, and promote sustainable agriculture.

Goal 3: Ensure healthy lives and promote well-being for all at all ages.

Goal 4: Ensure inclusive and equitable quality education and promote lifelong learning opportunities for all.

Goal 5: Achieve gender equality and empower all women and girls.

Goal 6: Ensure availability and sustainable management of water and sanitation for all.

Goal 7: Ensure access to affordable, reliable, sustainable, and modern energy for all.

Goal 8: Promote sustained, inclusive, and sustainable economic growth, full and productive employment and decent work for all.

Goal 9: Build resilient infrastructure, promote inclusive and sustainable industrialization, and foster innovation.

Goal 10: Reduce inequality within and among countries.

Goal 11: Make cities and human settlements inclusive, safe, resilient, and sustainable.

Goal 12: Ensure sustainable consumption and production patterns.

Goal 13: Take urgent action to combat climate change and its impacts.

Goal 14: Conserve and sustainably use the oceans, seas, and marine resources for sustainable development.

Goal 15: Protect, restore, and promote sustainable use of terrestrial ecosystems, sustainably manage forests, combat desertification, and halt and reverse land degradation and halt biodiversity loss.

Goal 16: Promote peaceful and inclusive societies for sustainable development, provide access to justice for all and build effective, accountable and inclusive institutions at all levels.

Goal 17: Strengthen the means of implementation and revitalize the global partnership for sustainable development (UN, 2015, p. 15–16).

In this respect, examining the toxicity of chemicals or elements used in PV production and reducing such usage can support this aim. Goal 7 of the SDGs aims to ensure access to energy and hence strongly integrates with the *energy efficiency* principle. Key targets for this SDG Goal include increasing renewables in global energy supplies, enhancing international cooperation in accessing clean energy, and promoting investments (Targets 7.2, 7.a). The *labor* principle strongly maps on to Goal 8 that aims to promote productive employment and "decent work" by helping to support international labor standards for worker rights and working conditions. Targets 8.7 and 8.8 of this SDG Goal refer to the need to "eradicate forced labor" and "protect labor rights and promote safe and secure working environments" (UN, 2015). By implementing this principle, the PV sector can support high quality job creation and wider economic development. In addition, the principles for waste reduction, climate emissions, and ecosystem protection all relate to specific SDG Goals and targets, particularly Goals 12, 13, and 15, respectively and their associated targets. Finally, the principle for eliminating *conflict mineral* use connects directly to the UN SDGs (Goal 16) that specifically promotes peaceful and inclusive societies that provide access to justice. Targets also are established for reducing violence (16.1), ending the exploitation of children (16.2), reducing illegal financial and arms flows (16.4), and preventing terrorism and crime (16.a).

5.4 PRACTICAL APPLICATION

These principles could be integrated into national governance, to guide future regulatory responses in developing countries. As described in Chapter 4, industry in European states is strongly regulated, with EU Directives and other policy instruments addressing potential impacts from these different themes. Similar federal and state level controls exist in the United States. However, this regulatory environment may be inducing a race to the bottom effect, as "today most PV modules are outsourced to and manufactured in non-OECD countries" such as China (Yue et al., 2014, p. 669). To avoid such inequalities, developing countries could seek to embed sustainable development much more into their domestic sectors, although the need for economic development per se may prove a significant barrier. Given the anarchy of the global international system, it may also be problematic to employ existing global trade regimes such as the World Trade Organization (WTO) to enforce better industry-led environmental and social protection in developing countries—an area where it has proved weak in the past (Connelly et al., 2012). In this case, the role of business itself should be encouraged.

One of the critical features of developing a sustainably green economy (eg, Barbier and Markandya, 2013) will be the active engagement of corporate actors. For some scholars of global governance (eg, Dingwerth and Pattberg, 2006), governing globally should involve nonstate actors such as multinational business, given its transboundary characteristics. Multiple examples of governing beyond the state are emerging globally (Risse, 2011). Yet, few binding regimes regulate the activities of MNCs. Global economic relations are largely framed by the WTO rules, which have often taken a deregulatory response to sustainability: leading to criticisms of its neoliberal agenda. India's attempts to expand its Jawaharlal Nehru National Solar Mission (see Chapter 4), to generate 20,000 MW of solar power by 2022, were challenged by the United States under WTO dispute resolution mechanisms, on the basis that the Mission plan discriminates against foreign companies by supporting indigenous Indian manufacturers through a domestic content requirement for subsidies (Rao, 2015). The rational economic self-interest of states, therefore, can provide a powerful disincentive to embedding sustainable development in global regimes, as IR theorists would no doubt predict. However, MNCs have the power to move and shape the global economic order in ways not readily available to national governments. Global MNCs now operate in transboundary geopolitical spaces and can exercise significant financial power, suggesting that their cooperation in developing a sustainable industry will be central in the future. In this respect, developing industry-led solutions will be vital.

5.4.1 Corporate Social Responsibility

As described previously, some national industry associations have started to support sustainable development initiatives, largely in the form of information provision and voluntary producer schemes. Individual businesses can also do more to enhance their sustainability credentials through voluntary measures. Although CSR[4] could be criticized for allowing business to pay lip service to this agenda, for many PV companies their CSR schemes are already a central mechanism for communicating their contribution to sustainable development to consumers and can be an important marketing device for securing custom. According to the European Commission (2001, p. 6), CSR is a "concept whereby companies integrate social and environmental concerns in their business operations and in their interaction with stakeholders on a voluntary basis." Information on such integration is then made publicly available by annual reporting. Many major solar companies are now establishing CSR strategies, influenced by emerging international norms developed by the UN, EU, and other bodies. For example, one leading manufacturer, Yingli, based in China, first published its Corporate Sustainability Report in 2013 showing how the company is integrating sustainability into its business practices (Yingli, 2013).

---

4. Also CSR is sometimes referred to as Corporate Sustainability Reporting, depending on the focus of the indicators used.

5.4.2   Environmental Management Systems

EMS are also widely employed by businesses worldwide, with two main approaches utilized. Firstly, the International Organization for Standardization (ISO) has introduced a suite of EMS standards under its ISO14001 series. Updated in 2015, ISO14001 provides a model for businesses to adopt environmental management throughout their operations. Secondly, ISO14001 can be compared with the European Union's Eco-Management and Audit Scheme (EMAS) which differs in respect of more stringent requirements for assessing environmental performance, employee involvement, registration, and the availability of information made public. Scope therefore exists for companies to adopt EMS approaches that integrate sustainability principles. The Japanese PV module manufacturer Mitsubishi Electric, for example, has gained ISO14001 certification for its Nakatsugawa Works in central Japan (Mitsubishi Electric, 2015). Solarcentury, a UK-based company that supplies solar technology in Europe, gained EMAS accreditation in 2008, publishing the results in an annual environmental statement (eg, Solarcentury, 2014). Some solar companies have adopted multiple designations, with SunPower, for example, implementing ISO14001 and OHSAS18001 (Occupational Health and Safety Assessment), in addition to engaging in sustainability reporting, carbon disclosure, supplier diversity, the elimination of conflict minerals in supply chains, and a voluntary take-back schemes for PV modules in Europe.

5.4.3   Life Cycle Assessment

Another industry-led approach that supports sustainability is LCA of PV products. Different definitions are evident globally but LCA is described by the ISO as:

> ... a technique for assessing the environmental aspects and potential impacts associated with a product, by - compiling an inventory of relevant inputs and outputs of a product system; - evaluating the potential environmental impacts associated with those inputs and outputs; - interpreting the results of the inventory analysis and impact assessment phases in relation to the objectives of the study. (ISO, 1997, p. iii)

According to the USEPA (2015), LCA aims to help manufacturers "make more informed decisions through a better understanding of the human health and environmental impacts of products, processes and activities." This technique works by allowing "for product comparison and strategic decision-making with regard to systemic inputs and outputs, as well as the development and incorporation of end-of-life design strategies" (Pryshlakivsky and Searcy, 2013, p. 115). In this respect, LCA is a decision support tool aimed at providing a "cradle-to-grave" analysis of impacts across the whole lifecycle (Pryshlakivsky and Searcy, 2013). Achieving this aim generally involves several steps, primarily through production of an initial inventory of inputs, evaluating potential environmental impacts, and finally interpreting the results to guide decision making (USEPA, 2015). Different systems of LCA are employed worldwide, although the most common is the ISO 14040 series (International Organization for Standardization, 1997).

In relation to the solar PV industry, LCA can help manufacturers identify and quantify environmental impacts throughout the production cycle, including final recycling or disposal of technologies. Assessments can also be integrated into EMSs to underpin wider sustainability of PV businesses. However, the majority of published LCA academic studies of PV technologies do not assess social and environmental impacts. Journal articles on LCA, as applied to PV technologies, suggest that most assessments focus on energy efficiency, EPBT and GHG emissions, that is, impacts that are relatively easy to quantify (Table 5.2). Few include other, less quantifiable, impacts such as toxicity of materials, human rights, biodiversity loss, and social effects, although some attempts have been made to develop broader sustainability indicators for LCA assessments of PV systems (eg, Traverso et al., 2012). Consequently, LCAs should adopt a broader conception of sustainability when applied to PVs, otherwise they can present only a partial representation of potential impacts for business and policy-makers.

TABLE 5.2 A Summary of Selected LCA Studies on PV Technologies and Impacts Assessed

Article	PV Technology	LCA Impacts Considered
Fthenakis and Kim (2007)	CdTe PV modules. These are compared to three crystalline Si-PV modules	Energy use, gaseous, and heavy metal emissions
Graebig et al. (2010)	Ground-mounted PVs	Nonrenewable energy input, greenhouse gas emissions (GHG), acidification, eutrophication, and soil erosion
Stoppato (2008)	Polycrystalline silicon PVs	Energy use, EPBT, and CO_2 mitigation potential (PCM)
Cucchiella and D'Andamo (2012)	Roof mounted PVs: monocrystalline silicon; polycrystalline silicon; amorphous silicon; cadmium telluride; copper indium gallium selenide	EPBT, GHG emissions/kWh, energy return on investment, greenhouse gas payback time (GPBT), greenhouse gas return on investment (GROI)

Continued

TABLE 5.2 A Summary of Selected LCA Studies on PV Technologies and Impacts Assessed—cont'd

Article	PV Technology	LCA Impacts Considered
Alsema et al. (2006)	c-Si, a-Si, Cu(InGa)Se2, CdTe	EPBT, GHG emissions + limited consideration of toxic emissions, resource supply, health and safety risks
Laleman et al. (2011)	3 kWp PV system	Cumulative energy demand, global warming potential, EPBT
Sherwani et al. (2010)	Monocrystalline, polycrystalline silicon	EPBT, GHG emissions
Garcia-Valverde et al. (2009)	4.2 kW p stand-alone photovoltaic system (SAPV)	EPBT, CO_2 emissions
Fthenakis and Kim (2011)	Monocrystalline silicon, multicrystalline, ribbon-silicon, cadmium telluride PV technologies	EPBT, environmental, and health impacts
Rocchetti and Beolchini (2015)	Copper indium gallium selenide (CIGS) and cadmium telluride (CdTe) thin-film photovoltaic panels	Abiotic depletion, acidification potential, eutrophication potential, global warming potential, ozone layer depletion potential, and photochemical ozone creation
Yue et al. (2014)	Monocrystalline silicon, multicrystalline silicon, ribbon-silicon	EPBT, energy return on investment, GHG emissions

LCA, life cycle assessment.

5.4.4 Extended Producer Responsibility

The solar PV sector could do more to support the reuse and recycling of end-of-life systems by industry, either through implementing regulations or industry-led responses. The potential benefits of recycling PV modules are well established (eg, Fthenakis, 2000). One potentially significant governance response is EPR. According to the OECD, EPR can be defined as an:

… approach under which producers are given a significant responsibility – financial and/or physical – for the treatment or disposal of post-consumer-

products. Assigning such responsibility could in principle provide incentives to prevent wastes at the source, promote product design for the environment, and support the achievement of public recycling and materials management goals. (OECD, 2015)

Chapter 3 noted that there are significant toxicological threats posed by the sector through the redundancy of technologies which then require disposal. In Europe, the EU WEEE (Waste Electrical and Electronic Equipment) Directive therefore compels a form of EPR that obliges manufacturers to organize collection and disposal systems for waste electronics. Yet, despite such regulation, a significant trade has emerged, both legal and illegal, between the EU and developing countries for waste electronic equipment. Such waste can be disposed of under poorly regulated and hazardous conditions in countries such as China and India (eg, Shinkuma and Managi, 2010).

In the United States, state legislation consists generally of bans on the disposal of hazardous waste in landfills. Such bans do nothing to reduce product toxicity or the total volume of consumer waste—they simply shift the burden of toxic waste collection and management to already under-funded local and regional governments. In California, for example, local governments spend a total of more than $100 million a year collecting and properly managing household hazardous products alone. However, US manufacturers are increasingly being forced by state laws to apply so-called EPR to their products. Californian local governments are working with the California Product Stewardship Council (http://www.calpsc.org/) to help move the state from a government—and rate-payer-funded—toxic waste disposal model to one that puts responsibility for product disposal on manufacturers. This EPR model not only reduces public costs, but also drives improvements in product design that promote environmental sustainability. Solar energy can play an important role in addressing climate change and also in stimulating the "green" economic sector. As the solar industry takes off, an EPR approach will be critical to ensuring the safe disposal of decommissioned products and also to ensuring that the costs of managing this new flood of e-waste do not fall on local governments and communities.

To date, more than 15 US states have passed EPR legislation, based on a variety of different policy models. California has also begun implementing its E-Waste Recycling Act (SB 20, passed in 2003), which adds a $6–$10 advance disposal fee to the purchase price of computer monitors and televisions to help fund e-waste collection and management efforts. Solar panels have not yet been covered, but electronics with similar toxic materials are currently regulated through this legislation. It is important to note that while the SB 20 fee supports the collection of televisions and computer monitors for recycling, it is not able to stop the export of California e-waste to developing countries that do not have the infrastructure to handle it. Another producer-responsibility approach requires brand owners to create a plan for collection and responsible recycling of their spent items (at no cost to local governments) before products can be sold within the regulating jurisdiction, such as a US state or Canadian province. This approach is being taken in Washington State, New York City, and British Columbia.

In Japan, the leading PV manufacturing country for most of the past decade, laws require the collection of electrical appliances (such as refrigerators and washing machines) for recycling, but computers and other electronics products are not specifically included. A fee is assessed on electrical appliance retailers, with retailers required to take back products and transport them to collection sites.

On a global scale, the OECD has consequently issued general guidance for EPR (OECD, 2001) for application in member states and by industry. Such programs potentially offer a number of advantages for sustainability in the PV sector, most notably reducing waste, removing toxic materials from the environment, and encouraging better product design. Indeed, the potential to "design out" waste impacts within a closed-loop or "circular" economy approach would present a significant move forward in how electronic equipment generally is produced worldwide.

5.5　SUMMARY

Previous chapters identified potential impacts from the PV sector and current governance approaches to countering them. What was apparent from the analysis was the lack of harmonization of national governance responses, suggesting a need for coordinated international-level action as this sector becomes ever more globalized. Problematically, the anarchy of the international political-economic system, combined with limited global mechanisms for integrating sustainability into transboundary economic activity, narrows the scope for a strong regulatory response between states. However, there are evident opportunities to integrate the development of the solar PV sector worldwide within the broader global sustainable development agenda, most notably the UN's SDGs. A framework of sustainability principles for the sector was therefore developed to link critical issues within the solar PV sector to this agenda. These principles could, in a normative sense, support national governments, particularly in developing countries, as they seek to expand their solar PV sectors. They can guide the introduction of regulatory mechanisms to control emissions, protect health and secure labor rights, in addition to informing instruments such as EIAs. More significantly, they could integrate with industry-led approaches to securing sectoral sustainable development by informing voluntary instrumental mechanisms such as corporate sustainability reporting, EMSs, LCAs, and EPR. Given the inherent problems of coordinating national-level sustainability, industry based approaches perhaps offer the best prospect for greater harmonization of sustainability in the global governance of the solar PV sector.

ABBREVIATIONS

c-Si	crystalline silicon
CdTe	cadmium telluride
CIGS	copper indium gallium selenide
CIS	copper indium selenide
CO_2	carbon dioxide
CSR	corporate social responsibility
EIA	environmental impact assessments
EICC	Electronic Industry Citizenship Coalition
EMAS	Eco-Management and Audit Scheme
EMS	environmental management systems
EPBT	energy payback time
EPR	extended producer responsibility
GHG	greenhouse gas
GROI	greenhouse gas return on investment
ILO	International Labour Organization
ISO	International Organization for Standardization
KPI	key performance indicator
LCA	life cycle assessment
MDGs	Millennium Development Goals
NO_x	nitrogen oxides
NO_2	nitrogen dioxide
PCM	CO_2 mitigation potential
SDGs	Sustainable Development Goals
SEIA	Solar Energy Industry Association
SO_2	silicon dioxide
SVTC	Silicon Valley Toxics Coalition
SWH	solar water heating
Te	tellurium
UNCED	United Nations Conference on Environment and Development
WTO	World Trade Organization

REFERENCES

Alsema, E.A., de Wild-Scholten, M.J., Fthenakis, V.M., 2006. Environmental impacts of PV electricity generation—a critical comparison of energy supply options. In: Presented at the 21st European Photovoltaic Solar Energy Conference, 4–8 September 2006, Dresden, Germany.

Baker, S., 2006. Sustainable Development. Routledge, Abingdon.

Barbier, E.B., Markandya, A., 2013. A New Blueprint for a Green Economy. Earthscan, London.

BBC, 2016. Where Many of Our Electronic Goods Go to Die. 8 January 2016, BBC, London. http://www.bbc.co.uk/news/business-35244018.

Cole, M.A., 2004. Trade, the pollution haven hypothesis and the environmental Kuznets curve: examining the linkages. Ecol. Econ. 48 (1), 71–81.

Connelly, J., Smith, G., Benson, D., Saunders, C., 2012. Politics and the Environment: From Theory to Practice. Routledge, London.

Cucchiella, F., D'Adamo, I., 2012. Estimation of the energetic and environmental impacts of a roof-mounted building-integrated photovoltaic systems. Renew. Sust. Energ. Rev. 16 (7), 5245–5259.

Dingwerth, K., Pattberg, P., 2006. Global governance as a perspective on world politics. Global Governance 12 (2), 185–203.

Electronic Industry Citizenship Coalition (EICC), 2014. Code of Conduct. EICC, Alexandria, VA. http://www.eiccoalition.org/standards/code-of-conduct/.

European Commission, 2001. Green paper promoting a European framework for corporate social responsibility. COM(2001) 366 final. Commission of the European Communities, Brussels.

European Photovoltaic Industry Association (EPIA), 2015a. Fact sheets. http://www.epia.org/news/fact-sheets/.

European Photovoltaic Industry Association (EPIA), 2015b. The Water Footprint. EPIA, Brussels. http://www.epia.org/news/fact-sheets/.

First Solar, 2015. Social Impact: First Solar Conflict Minerals Policy. First Solar, Tempe, AZ.http://www.firstsolar.com/en/About-Us/Corporate-Responsibility/Social-Impact.aspx.

Fthenakis, V.M., 2000. End-of-life management and recycling of PV modules. Energ. Policy 28, 1051–1058.

Fthenakis, V.M., Kim, H.C., 2007. CdTe photovoltaics: life cycle environmental profile and comparisons. Thin Solid Films 515, 5961–5963.

Fthenakis, V.M., Kim, H.C., 2011. Photovoltaics: life-cycle analyses. Sol. Energy 85, 1609–1628.

Garcia-Valverde, R., Miguel, C., Martinez-Bejar, R., Urbina, A., 2009. Life cycle assessment study of a 4.2 kWp stand-alone photovoltaic system. Sol. Energy 83, 1434–1445.

Graebig, M., Bringezu, S., Fenner, R., 2010. Comparative analysis of environmental impacts of maize–biogas and photovoltaics on a land use basis. Sol. Energy 84, 1255–1263.

Han, J., Mol, A.P.J., Lu, Y., 2010. Solar water heaters in China: a new day dawning. Energ. Policy 38, 383–391.

Hasenclever, A., Mayer, P., Rittberger, V., 1997. Theories of International Regimes. Cambridge University Press, Cambridge.

Hosenuzzaman, M., Rahim, N.A., Selvaraj, J., Hasanuzzaman, M., Malek, A.B.M.A., Nahar, A., 2015. Global prospects, progress, policies, and environmental impact of solar photovoltaic power generation. Renew. Sust. Energ. Rev. 41, 284–297.

International Labour Organization (ILO), 2015. ILO Declaration on Fundamental Principles and Rights at Work. ILO, Geneva.

International Organization for Standardization (ISO), 1997. ISO 14040: Environmental Management—Life Cycle Assessment—Principles and Framework. ISO, Geneva.

Krasner, S.D., 1983. Structural causes and consequences: regimes as intervening variables. In: Krasner, S.D. (Ed.), International Regimes. Cornell University Press, Ithaca, NY.

Lafferty, W.M., Meadowcroft, J., 2000. Implementing Sustainable Development: Strategies and Initiatives in High Consumption Societies. Oxford University Press, Oxford.

Laleman, R., Albrecht, J., Dewulf, J., 2011. Life cycle analysis to estimate the environmental impact of residential photovoltaic systems in regions with a low solar irradiation. Renew. Sust. Energ. Rev. 15, 267–281.

Massey, D.B., 1995. Spatial Divisions of Labor: Social Structures and the Geography of Production, second ed. Routledge, New York.

Mitsubishi Electric, 2015. Solar Power: Overview. Mitsubishi Electric, Tokyo.http://www.mitsubishielectric.com/bu/solar/overview/pvplant.html.

OECD, 2001. Extended Producer Responsibility: A Guidance Manual for Governments. OECD, Paris.

OECD, 2015. Extended Producer Responsibility. OECD, Paris.http://www.oecd.org/env/tools-evaluation/extendedproducerresponsibility.htm.

Pierre, J., 2013. Globalization and Governance. Edward Elgar, Cheltenham.

Pryshlakivsky, J., Searcy, C., 2013. Fifteen Years of ISO 14040: A Review. J. Clean. Prod. 57, 115–123.

Rao, K., 2015. India's grand solar plans threatened by ugly US trade spat. The Guardian. 23 April 2015.

Revesz, R.L., 1992. Rehabilitating interstate competition: rethinking the race-to-the-bottom rationale for federal environmental regulation. NYUL Rev. 67, 1210.

Risse, T. (Ed.), 2011. Governance Without a State?. Columbia University Press, New York.

Rocchetti, L., Beolchini, F., 2015. Recovery of valuable materials from end-of-life thin-film photovoltaic panels: environmental impact assessment of different management options. J. Clean. Prod. 89, 59–64.

Sands, P., Peel, J., 2012. Principles of International Environmental Law. Cambridge University Press, Cambridge.

Securities and Exchange Commission, 2012. Final Rule: Conflict Minerals—SEC. SEC, Washington, DC.www.sec.gov/rules/final/2012/34-67716.pdf.

Sherwani, A.F., Usmani, J.A., Varun, 2010. Life cycle assessment of solar PV based electricity generation systems: a review. Renew. Sust. Energ. Rev. 14, 540–544.

Shinkuma, T., Managi, S., 2010. On the effectiveness of a license scheme for e-waste recycling: the challenge of China and India. Environ. Impact Assess. Rev. 30 (4), 262–267.

Silicon Valley Toxics Coalition (SVTC), 2009. Towards a Just and Sustainable Solar Industry. SVTC, San Francisco, CA.

Silicon Valley Toxics Coalition (SVTC), 2015. 2015 Solar Scorecard. SVTC, San Francisco, CA.

Solar Energy Industries Association (SEIA), 2012. Solar Industry Commitment to Environmental & Social Responsibility: Participant Handbook. SEIA, Washington, DC.

Solarcentury, 2014. Environmental Statement 2013–2014. Solarcentury, London.

Stoppato, A., 2008. Life cycle assessment of photovoltaic electricity generation. Energy 33, 224–232.

Strange, S., 1996. The Retreat of the State: The Diffusion of Power in the World Economy. Cambridge University Press, Cambridge.

SunPower, 2015. Conflict Minerals Policy. SunPower, San Jose, CA.http://us.sunpower.com/conflict-minerals/.

Traverso, M., Asdrubali, F., Francia, A., Finkbeiner, M., 2012. Towards life cycle sustainability assessment: an implementation to photovoltaic modules. Int. J. Life Cycle Assess. 17, 1068–1079.

United Nations (UN), 1948. The UN Declaration of Human Rights. UN, New York.

United Nations (UN), 2015. Transforming Our World: The 2030 Agenda for Sustainable Development. UN, New York.http://sustainabledevelopment.un.org/post2015/transformingourworld.

USEPA, 2015. Utility Resources: EPA's Life-Cycle Assessment Website. EPA, Washington, DC. http://www3.epa.gov/epa/conserve/smm/waterwise/wrr/util-res.htm.

Weiss, L., 1999. Globalization and national governance: antimony or interdependence? Rev. Int. Stud. 25, 59–88.

World Commission on Environment and Development, 1987. Our Common Future. Oxford University Press, Oxford.

Yingli, 2013. Yingli Green Energy: 2013 Corporate Sustainability Report. Yingli, Baoding.

Yue, D., You, F., Darling, S.B., 2014. Domestic and overseas manufacturing scenarios of silicon-based photovoltaics: life cycle energy and environmental comparative analysis. Sol. Energy 105, 669–678.

Chapter 6

Future Issues and Recommendations

6.1 INTRODUCTION

The photovoltaic (PV) industry has seen rapid growth with 135 GW of cumulative solar PV capacity across the world in 2013 (IEA, 2014). Silicon based PV modules are still ruling the market. Silicon is an important material for today's electronic market, which is growing linearly. It can offer more stable electronic or semiconductor devices, which attract many industries to invest. However, the processing of this material from silica to a usable semiconductor involves a tedious toxic-chemical process. The other PV materials, such as amorphous silicon, $Cu(In,Ga)Se_2$, cadmium telluride, organic solar cells, perovskite solar cells, and third generation solar cells also involve toxic-chemical usage in processing and cell manufacturing. This rapid growth has caused the PV industry to adopt and improve safety and environmental methods—starting with material processing, and on to the recycling of PV panels. The following recommendations can be implemented in PV materials processing and in the recycling of PV modules.

6.2 RECOMMENDATIONS

6.2.1 Materials Processing

Materials processing is an intensive process to produce an effective solar absorber with right band gap. The process starts with the mining and separating the minerals, handling the waste, cleaning the mineral, processing with chemicals, etc. In each step of the processing, the industries can reduce and eliminate things that affect the environment, human health, and pollute the air.

1. *Green mining*: It adopts best practices in mining processes to reduce the environmental impacts associated with the extraction and processing of metals and minerals. This has huge impact in selecting mining's ability to reduce its ecological footprint. It positively contributes to the reduction of chemical use, green house gas reduction, and better energy efficiency.
2. *Green chemistry*: It takes a life-cycle approach that involves each step of the process, such as raw-material extracting to product end-life. There

Solar Photovoltaic Technology Production. http://dx.doi.org/10.1016/B978-0-12-802953-4.00006-8

are defined principles of green chemistry including precautionary and pollution prevention measures to incorporate safer chemicals, minimum energy use, and less waste (Anastas and Warner, 1998).
3. *Environmentally friendly materials* reduce and eliminate the use of toxic materials in each processing step and develop a sustainable, green approach. The other way to tackle the use of toxic materials in PV cell manufacturing is to invent environmentally friendly materials.

6.2.2 Health, Safety, and Environment

Global PV industrial worker health and safety should be protected through the green jobs principles. Human contact to toxic chemicals should be regulated to global PV industry. More automated processes, with less human interaction, can be introduced in the materials process. Safety and environmental awareness in the handling of chemicals, mining minerals, engineering aspects in the industry supply chains, and end-life recycling should be created. The awareness of the effect of solar farm issues, such as visual impact, insect false focus impact, and the biodiversity effect in huge solar farms needs careful consideration.

6.2.3 Policies and Regulations

Chemicals such as cadmium, lead, mercury, brominated flame-retardants, and chromium are considered highly toxic and are associated with health and environmental impacts. Strong policies to regulate the use and handling of these chemicals should be put forward by concerned governments. Developing alternative methods and materials should be encouraged by industries to avoid toxic-chemical usage and for energy-intensive processes for semiconductor materials processing. The potential negative effects from chemicals, such as sulfur hexafluoride (SF6), need to be controlled through proper regulations. This precautionary approach should be achieved through incorporation of policies.

6.3 CONCLUSIONS

Finally, this book suggests that the global governance of solar PV impacts could be better coordinated around the notion of sustainability between national contexts. As Chapter 4 shows, while different countries have proved successful in promoting the growth of the solar PV industry through different policy mechanisms, there is variability in how the associated impacts are being governed. Given the transnational nature of this industry, where supply and demand for solar technologies are becoming increasingly separated between national contexts, better coherence on sustainable development would seem advisable. In this respect, Chapter 5 examined the scope for integrating sustainability within the sector using voluntary policy instruments such as Extended Produce Responsibility and Environmental Management Systems. A suite of sustainable development principles were derived from current industry sustainability guidelines and our analysis of its impacts (Chapter 3), with linkage made to the UN's Sustainable Development Goals. While by no means exhaustive, these principles could then support both industry-led voluntary initiatives and government policy worldwide in order to counter emerging problems from the solar PV sector.

However, research into the environmental impacts of solar PV technologies is still ongoing, and there has been only limited discussion of governance implications. Although our research is intended to address these gaps in the literature, there are still areas where further research is evidently required. Firstly, as Chapter 3 demonstrates our knowledge of the toxicological impacts of hazardous substances employed in PV technologies is still emerging. Future research could then focus on how risk from these substances is being quantified. Secondly, the issue of waste electronic material associated with solar PV is barely discussed in the literature, but given the life span of solar panels, may start to become problematic in future years. Safe-disposal options and the recycling of materials could become the object of valuable future research. Thirdly, the "social" impact from the solar industry warrants further attention by researchers. For example, while the issue of "conflict" diamonds has generated much media and political attention, the use of minerals from specific conflict zones in solar PV manufacture is not greatly understood by policy analysts. Lastly, further research at the interface between science and policy is required. Chapter 4 shows that very little comparative research has been conducted into how different national contexts govern the solar PV sector. While the chapter compares governance in five leading states, the industry is expanding rapidly on a global scale, raising questions over how such technologies are diffusing between other states, how these states are learning from each other, what governance responses are being adopted, and, significantly, what we can learn from governance "best practices." Answering these questions is timely and will contribute to broader debates over sustainable development and the creation of a global green economy.

REFERENCES

Anastas, P., Warner, J., 1998. Green Chemistry: Theory and Practice. Oxford University Press, New York, NY, p. 30. By permission of Oxford University Press.

IEA, 2014. Technology Roadmap: Solar Photovoltaic Energy. International Energy Agency, Paris.

Index

Note: Page numbers followed by *f* indicate figures, and *t* indicate tables.

Printed in the United States
By Bookmasters